I0047962

William Allan

Rough Castings in Scotch and English Metals

William Allan

Rough Castings in Scotch and English Metals

ISBN/EAN: 9783744730693

Printed in Europe, USA, Canada, Australia, Japan

Cover: Foto ©berggeist007 / pixelio.de

More available books at **www.hansebooks.com**

ROUGH CASTINGS

IN

SCOTCH AND ENGLISH METAL

BY

WILLIAM ALLAN

All day building compound Engines,
Is the glory of my time ;
But at nights,—I, with a vengeance,
Mould my thoughts with tools of rhyme.

Discordant notes a man doth bring,
When harpin' aye upon ae string.

LONDON : SIMPKIN, MARSHALL, & Co
EDINBURGH : OLIVER & BOYD

1872

LOAN STACK

SUNDERLAND :

PRINTED BY WILLIAM HENRY HILLS.

THESE

ROUGH CASTINGS

ARE INSCRIBED

TO ONE OF THE OLDEST ENGINEERS IN SCOTLAND

NAMELY

MY FATHER

PREFACE

AT the solicitations of a few valued friends I am induced to put before the Public a few samples of the last twelvemonth's Evening Castings from my mental cupola.

In doing so, I would have every purchaser remember, that by reason of the onerous duties and responsibilities of my daily profession—requiring a great amount of thought, and giving a great amount of care—I have not even " dressed " these castings, much less machined or fitted them properly together ; consequently, they go forth a " rough job," "just out of the sand," and require, I fear, to be put into the " vice," and touched up, with the " smooth file."

However, the casting of them has afforded the moulder many a foot of pleasure, and he trusts their inspection will afford, at least, an inch of delight to every one into whose possession they may come.

To the " Dressers" of such Castings, or detectors of blown holes, &c., I may say (without at all calling in question the skilful manipulation of their tools) that it would suit the moulder, if they used their bluntest chisels and lightest hammers in knocking off any " fins " or projections.

WILL O' THE WISP.

CONTENTS

X. *CONTENTS*

CONTENTS

MY PLEA FOR RHYMING

I AFTEN speir hoo I could fash
To spen' my time in rhymin' trash,
An' steek Scotch words wi' twaddlin' clash,
  An' think it's rhyme ;
But thochts sometimes come wi' a rush,
  An' verses chime.

Tho' many a year's gane owre my head,
Sin' I gaed to the engine trade,
Sang-makin' to me ne'er was laid,
  'Twas 'yont my range,
But noo whan gettin' auld and staid,
  I try the change.

A towmond yet has nae gane by,
Sin' I my 'prentice han' did try,
The feck o' them afore ye lie
        In this sma' volum',
Dootless great fauts ye will espy,
        In ilka colum'.

Tho' lang frae Scotland I ha'e been,
An' mony countries ha'e I seen
(Wi' truth I tell't, I've tried a wheen),
        My ain's the chief.
Auld Scotland stan's, wi' glorious sheen,
        In bauld relief.

Scotland! what heart-grip's in thy name,
That ither countries canna claim!
What Scottish tongue daur not proclaim
        His country's worth?
A Scotchman's boast is Scotland's fame,
        Owre a' the earth.

See Scotchman in a foreign land,
Cared for by nane, by a' unkenned,
His heart leaps wi' exultin' bound
    At mither-tongue ;
The tearfu' e'e—the weel-grip't hand—
    An' a' unstrung.

Sae when I'm dune wi' business hours,
An' plagued in mind wi' business sours,
I coort my Scotch poetic powers,
    An' bring a calm.
The bonnie queen min's fire devours,
    Wi' soothin' balm.

Wi' pen in han', an' pipe in cheek
(The words come fast in clouds o' reek),
Nae greater pleasure dae I seek
    Then my ain hoose,
Jean by my side, thrang wi' her steek,
    An' a' sae crouse.

Tho' warld's wealth or grandeur flee,
Tho' backs of freen's I only see,
Tho' reft o' a'—in poverty,
    Pitied by nane,
The love an' sel's enough for me,
    O' my ain Jean.

Fu' aft, this way my time at nicht,
I spen' in hame-owre great delicht;
The pleasure o' the Muse's flicht
    Sweet joy affords.
She aft mak's dull time unco licht,
    Wi' kindly words.

My Muse aftimes wi' joy appearin',
Wad darken thochts o' engineerin',
Into her arms I gang careerin';
    I canna haud;
Verse after verse sae quickly rearin',
    To please the jaud.

She'll saftly whisper in my ear,
Sing thou o' Scotland loved an' dear;
Thy verses Scotia's impress bear
      In Scottish rhyme,
The lan' o' men whose deeds aye are
      Supports o' Time.

An' aft it gi'es the cuttie pleasure
To coax me into English measure;
Frae oot o' baith, notes at my leisure
      I aften mak'.
A nation's rhyme's a nation's treasure
      In mither talk.

Critics may sneer wi' anger fierce,
Or pu' the skin frae aff my verse,
An' lea' the banes o' flesh gey scarce,
      I dinna care;
Thae paper hawks book's hearts sune pierce—
      They dinna spare.

The notes o' nightingales may soun',
The cheep o' sparrow far aboon
Tho' ane is hardly ca'd a tune,
      Ye're no to kill't ;
The effort wi' its best is dune,
      There's that intil't.

Sae after a', I dae nae wrang
Gin aiblins I should mak' a sang ;
The sin's my ain, sae haud yer tongue,
      Ye critic chiels.
Sen' me, but no my rhyme, amang
      Tormentin' deils.

HARK! 'TIS THE PIBROCH

COME to the mountains, where grows the brown heather,
  Come to the brackens, where lie the red deer,
Come to the valleys, where stalwart men gather,
  And whose bonnie daughters we aye will revere.
Ours is the land famed in song and in story,
  Where hearts strong and true aye unconquered will
    stand,
Whose pride is to tell of their forefathers' glory,
  Who fought for their chieftains and bonnie Scotland.
    Hark! 'tis the pibroch played, down in yon
      lonely glade—
    Strains that have stirred oft bold hearts under-
      neath,
    And while the heather waves over their honoured
      graves,
    Scotsmen as dauntless will tread o'er the heath.

Come to the land that has baffled invasion,
  Each mountain and glen has famed story to tell,
How tartan-clad heroes, the pride of their nation,
  With broadsword and targe the strong foe did repel.
We have no craven hearts reared on our heather,
  But nursed 'mid the cold mists with no gentle hand,
Our mountain-bred valour with us ne'er will wither,
  And deeds done before we can do for Scotland.
    Hark! 'tis the pibroch, &c.

Our forefathers' battles with pleasure recounting,
  Unsullied shall shine on the long scroll of fame;
And long as the tartan our breasts is surmounting,
  So long shall their spirits dwell in us the same.
Ours is the land of brave men never vanquished,
  Who dare, do, or die at their chieftain's command;
We all love the country whose name is untarnished;
  Hurrah for the mountains of bonnie Scotland!
    Hark! 'tis the pibroch, &c.

DARKENED HOURS

WE may have days of brightness long,
  And ne'er see sky o'ercast ;
The sea may sing its peaceful song,
  Unruffled by a blast.
The night may shine with starlit glow,
  The moon undimmed may rise,
The woods reply to night-breeze low,
  That smiles in cloudless skies.
    Thus lives may have a sunny side,
      And paths be strewn with flowers,
    Some clouds may yet our hopes divide,
      And bring us darkened hours.

For skies have oft, in anger wild,
  Obscured a brightening ray,

And ocean sleeping as a child,
  Awakes to wild wind's play.
Aye, tempest-torn has been the day
  By strong gales mighty power,
And dark clouds oft have paled the ray
  At end of sunlit hour.
        Thus lives may have a sunny side, &c.

Our lives may be undimmed, serene,
  Thro' many happy years,
A pall may yet o'erspread the scene,
  And bring us bitter tears.
We may with skill have plans arranged,
  And think this world is ours,
But sunny moments oft have changed,
  And left us darkened hours.
        Thus lives may have a sunny side, &c.

GIE ME A HEART

WHAT guid is the siller, or constant life-fechtin',
   Gin peace to your mind ye hae nane?
The want o' some fond heart, a' dreich cares to lichten,
   Maks life an absurdity vain.
That chield is a coof wha but lives for himself,
   An' lasses looks on wi' disdain.
The scrapin' an' hoardin' his heart-freezin' pelf,
   Is a' his sma' mind can contain.
     But gie me a heart keepin' time wi' my ain,
       An' gie me the bricht, lauchin' ee,
     An' gie me love's saft han's, whan writhin' we' pain,
       Drap owre me love's tears whan I dee.

That wealth can gie comforts, it's nae use denyin',
   But lanely they're naked an' pared,

Bereft o' their beauty, wha wad them be buyin'?
    Life's pleasures are true whan they're shared.
Sae wealth without love is a braw ship unhelm'd,
    An' aft when adversities blaw,
On life's rocks fast drivin', a' sune is o'erwhelm'd,
    Nane pities the loss or doonfa'.
        Sae gie me a heart keepin' time wi' my ain, &c.

Life's rough ways are levelled whar love is abidin';
    Then wealth maks the pleasure aye mair;
Love growin' frae love, in twa hearts aye confidin',
    Strips poverty's terrors fu' bare.
I care na to live gin I hae na the smiles,
    O' some ane whas lovin' heart shows
The bricht side of life, an' a' dark oors beguiles,
    Wi' the licht that kens na repose.
        Sae gie me this heart keepin' time wi' my ain,
        &c.

Oor happiness, finished by Powers above
    Sune after the warld began,

A' meant to be cantie, an' live aye in love,
  An' work oot a pairt o' the plan.
Sae fouk ne'er are happy whan livin' alane;
  There is aye a something ajee;
The heart only echoed it's ain wa's within
  Life's beauties an' worth ne'er can see.
    I'll loe aye ae heart beatin' time wi' my ain,
      I'll loe aye the bricht lauchin' ee,
    Saft, saft is love's pillow whan writhin' wi' pain,
      Sair, sair are love's tears whan ye dee.

FAREWEEL TO THE SNAW

AINCE mair dreary Winter's departed, unmourned,
An' Spring wi' its sweet voice again has returned,
Wi' clear skies, an saft win's that cheerily blaw,
Aye singin' sae blithely, Fareweel to the snaw.
　　　　Birds merrily cheepin',
　　　　Buds timidly peepin'.
A life-breathin' lilt is, Fareweel to the snaw.

The hill-taps divestit o' mantle o' white,
Wi' joy kiss the clouds in their youthfu' delight.
The valleys wi' verdure are decked oot fu' braw ;
Ilk blade nods sae joyous, Fareweel to the snaw.
　　　　Flowers bonnily bloomin',
　　　　The saft breeze perfumin'.
A life-breathin' lilt is, Fareweel to the snaw.

Thus Nature sae queenly, wi' love-bearin' smiles,
Awakes in her beauty, an' a' hearts beguiles;
Frae gems o' the fairest, her dew draps doon fa',
A' whisp'rin' sae clearly, Fareweel to the snaw.

    Draps tenderly hingin'
    Frae gems that are singin'.
A life-breathin' lilt is, Fareweel to the snaw.

Whan life's dowie winter brings death's chills aroun',
Nae spring comes to cheer us wi' life-givin' soun';
To hame—'neath the greensward, we a' wear awa,
Nae mair to sing fondly, Fareweel to the snaw.

    Freens sorrowfu' leavin'
    'Mid sabbin' an' grievin'
We nae mair sing fondly, Fareweel to the snaw.

THE OLD WINDMILL

'Twas here oft I sat on the old miller's knee,
Who fondly caressed me when I was a child;
'Twas here, in youth's heyday, I sported in glee,
And the long happy hours of summer beguiled.
The shadow of bygones around me is clinging,
Bright visions appearing that thro' my heart thrill,
And tears of old age with the loved past are mingling,
As I sit by the wreck of the Old Windmill.
    Ah me! all is changed, and my dim eyes fill,
    Recalling the past in the Old Windmill.

Oft, when afar from the scenes of my childhood,
I longed to revisit that home once again,
Dreams of the green spot whereon the old mill stood
To slumbers gave pleasure, to wak'ning heart pain.

Thro' my long wand'rings one sole wish I carried,
That oft darkest moments with hope would instil—
When ended life's battling, that I might be buried
'Neath the green grass that waves round the Old
  Windmill.

  Ah me! all is changed, and my dim eyes fill,
  Recalling the past in the Old Windmill.

Alack! I behold thee to ruin fast falling,
Thy silence gives sorrow and pain to the sigh,
On loved ones departed I vainly am calling,
And rooks from their crannies but give me reply;
The breezes are blowing; birds warbling, rejoice,
And blooming in beauty the meadows are still;
But sadly I miss now the miller's kind voice
That welcomed me first to the Old Windmill.

  Ah me! all is changed, and my dim eyes fill,
  Recalling the past in the Old Windmill.

Back to the old home I weary have wandered,
'Tis but as a stranger I linger around ;
Gone are companions on whom oft I've pondered,
No welcomes of childhood are now to be found ;
Hard by the old miller in death now is sleeping,
His mound in the churchyard I see on yon hill ;
That soon I'll be near him, no more to be weeping,
Is silently told by the Old Windmill.

    Ah me, happy change, where no tears can fill
    The eyes now beholding the loved Old Mill.

COCKEN WOODS ARE BONNIE

THE shadows o' evenin' owre Cocken were creepin',
   An' blackbirds were warblin' their notes lane an' clear;
The ruins o' Finchale their lane watch were keepin',
   And rumlie an' low was the sang o' oor Wear.
As lanely I sauntered, the fair scene enjoyin',
   A birdie frae bush tap the brackens amang,
Gae lilt true and saftenin', my heart a' decoyin',
   Enraptured I listened to its bonnie sang.

SONG OF THE BIRDIE.

I hae come frae a fair south'ren clime,
   Where orange groves bonnilie bloom,
I hae sang on sierras sublime,
   Where zephyrs are charged wi' perfume,
I hae warbled 'mang rich growin' vines,
   An' aft by some clear stream fand rest,

Tho' in glades 'o maist flowery designs,
　I ne'er saw a nook for a nest.
　　　But Cocken woods are bonnie,
　　　　An' Cocken dells are fair,
　　　Brighter far than ony,
　　　　Nane wi' them compare.

I hae sought repose in sunnier lands,
　Where scenes maist enchantin' abound,
An' the blue seas roll on golden strands,
　Alas! 'twas not there to be found.
But a breeze frac the far northern skies
　Aince ruffled my longin' wee breast,
　Whisp'rin' saftly, Come fly where I rise,
I ken o' a nook for a nest,
　　　In Cocken woods sae bonnie,
　　　　An' Cocken dells sae fair,
　　　Brighter far than ony,
　　　　Nane wi' them compare.

Its liltie sae lovely the wee birdie endit,
   And flutt'rin a wee, flew sae blithely awa ;
Responsive my thochts a' in unison blendit,
   For fairer than Finchale I ken nane e'er saw.
Tho' lands may be famous wi' mountains and valleys,
   And ruins historic fu' grandly appear,
Yet Nature at Cocken has made her ain palace,
   Sae seek not for beauty, while flows on the Wear.

THE HEART-GRIP IS AYE THERE

LAT ither poets sing in themes,
Frae lan's o' sunny skies—
'Bout love-sick chiels, aye drunk wi' dreams
O' beauties wi' dark eyes,—
Fine words may gild but winna bring,
The feelin' unco rare,
Tho' voices rich, their sangs may sing,
The heart-grip is na there.
 But Scotland's sangs, in Scotland's tongue,
 Some spell aroun' them bear,
 See Scottish sangs, where'er they're sung,
 The heart-grip is aye there.

Sangs may be like the driftit snaw,
Sae bonnie, clear, and bricht,

A wee time white, but melt awa,

An' sune are lost to sicht ;

Awa wi' a' your snawy verse,

Its notes sune oot aye wear,

Stan' oot ! the sangs Time canna pierce,

The heart-grip is aye there.

　　'Tis Scotland's sangs, in Scotland's tongue,&c.

Wi' firmest base, oor country's lays,

Deep in oor hearts aye lie,

An' Scotland's sangs, wha daurna praise,

They wither not, nor die ;

Oor mountain rhyme a', a' can sing,

The notes nae wershness bear,

The smile or tear they aye can bring,

An' heart-grip is aye there.

　　'Tis Scotland's sangs, in Scotland's tongue,&c.

ADVICE TO YOUTH

My son, as thro' this warl' ye gae,
  Aye bear these facts in mind,—
Let fouk that's speakin' hae their say,
  An' to their fauts be blind.
Hear a' that's said wi' earnestness,
  An' tho' it should be wrong,
Ne'er gar them think their sense is less,
  By blawin' up your own.

Whate'er remarks ye hae to gie,
  Let calm persuasion lead.
Mind tell the truth, aye shun the lee,
  Then fouk will gie ye heed ;
But be na blate, adorn the richt,
  An' fear na to say No ;
Ne'er mak yoursel' a snoovlin' wicht,
  Far rather freens forego.

Aboon a' mind that temper's aft
   The life o' mony spoiled ;
Ne'er angry get, it shows you're saft,
   Men treat ye as a child.
Ah man ! I aft hae dearly paid
   For lack o' temper's reins,
When anger reamed, then words were said
   That still hae left their stains.

Whate'er your walk in life may be,
   Aye tread it wi' strong will ;
Mind, should ye uphill wark e'er flee,
   You're at the bottom still.
Think na that success will attend
   The sweat deplorin' man ;
Think na, to work out life's best end,
   By wondrin' "If you can."

Na, na, it winna suit the age,
   Ye wad be left ahin' ;
"I will" maun be your maxim sage,
   Stick to't thro' thick and thin,

An' ne'er forget, should ye be placed
    Your fellow-men aboon,
That justice' brow is nobly graced
    When a' get justice dune.

Whate'er ye say, whate'er ye do,
    Ne'er breathe the would-be "I";
Ne'er tak' ye credit, tho' it's due,
    Your worth let ithers spy.
Aye guard against self-mightiness,
    For men mak' this a rule,
Before ye, aye to acquiesce,
    Behind, think ye a fule.

Ye ken temptations come unsought
    Wi' mony dazzlin' train;
Their momentary pleasures bought
    Are paid wi' years o' pain.
Be proof to a' mind-wreckin' snares,
    Depravin' thochts aye flee;
As your companions, sae your cares,
    An' sae your life will be.

I LEFT ONE LOVING HEART BEHIND

I LEFT one loving heart behind,
   In sorrow and in tears,
A heart to anxious hope consigned,
   O'ershrouding life with fears.
Stern fate had marked me for its prey,
   And tore me from her side,
The voice of love would whisper "stay,"
But duty's louder cried.

      I left one loving heart behind,
         In sorrow and in pain,
         Love's strongest bands will yet us bind,
         When I return again.

Tho' trammelled with no passions vain,
   Our fairest pleasures flow,

'Tis love that gives all parting pain,
　We shrink beneath its blow.
Her love-fraught tears, the longing kiss,
　With grief my heart did fill;
The parting moment wrecks our bliss,
　Its sorrow haunts me still.

　　　I left one loving heart behind, &c.

But know, my Jean, tho' weary seems
　The anxious hope-charged hours,
Thy lovely form will fill my dreams,
　And mine may be in yours.
By Him who dwells in realms above,
　By manhood's brightest claim.
I'll bring ye back unsullied love,
　E'en purer be its flame.

　　　Tho' leaving thee in tears behind, &c.

MAN OF THE AGE

'Tis the age of pushing, driving,
  Lives are naught when gold's at stake ;
To be wealthy, how contriving,
  Makes the man all else forsake.

Toiling, moiling, life a-spoiling,
  Never thinking gain is loss,
And his better nature spoiling,
  In exchange for fleeting dross.

To be rich is the sole prayer,
  Oft its getting manhood drowns,
Ambition's whispers are a snare,
  Magnet smiles oft hide dark frowns.

Small flowers may give drops of honey,
　　Scrape and gather for your hive,
Heed not conscience—get the money!
　　Weighty purses best do thrive.

This the object of your being,
　　This the only god and goal;
Vot'ry blind the future fleeing,
　　Would for gold e'en sell his soul.

Comes some winter you consuming,
　　Pound of gold for grain of health,
Death your sordid hive perfuming,
　　Grants some spendthrift all your wealth.

'Tis the age of pushing, driving,
　　Man will never know content,
To be wealthy, how contriving.
　　Heart enshrined in cent. per cent.

THOUGHTS ON STEAM

LOVED Charmer of my labour-ended hours,
   Oh lyre! yet once, I fain would touch thy strings;
Tho' blushing Hope within me trembling cow'rs,
   The impulse, all pervading, upward springs,
Tipping with light my fancy's flutt'ring wings :
   Again, my lyre, around me weave thy spell,
Let dulcet notes with sweetest murmurings
   Reverberate, and onward me impel,
   To sweep thy chords aright, and visions vain dispel.

I seek not other lands whose classic past
   Affords, full oft, meet off'rings to the lyre,
Nor where traditions hoary shadows cast,
   That fill the vot'ry with poetic fire,
Their silent grandeur fanning his desire.
   Where shall my fancy find a fitting theme,

If Past can not, the Present must inspire ;
   Come, then, my muse, I'll sing no misty dream,
   But trace in numbers rude the path of Mighty Steam.

The records of the past to us reveal
   How Alexandrian Hero, with great joy
Elated was, when first his simple wheel
   By Steam revolved, a noisy whirling toy,
To wile the dullard hours he would employ.
   Dream't not the gaping crowd the germ was there,
Of power unsurpassed, that would destroy
   Barbaric gloom, and in its march would bear
   Dark Bigotry's death knell, and Bethle'm's Light
     declare.

Knew not the haughty Romans, as they gazed,
   That monuments far greater than they reared
Would by this " hissing thing" one day be raised,
   While their proud works had all but disappeared
From off the earth, where they as conq'rors steered.

Ah! recked they not the little pow'r thus shown,
Could ne'er by withering hand of Time be seared,
    Had it been grasped, perchance, their tow'ring throne
And empires that they ruled might still have been
        their own.

'Twas not to be ;—'mong all their gifted men,
    The Giant's birth ne'er gave a passing thought ;
War, glory, empire was their highest ken,
    For which their conq'ring legions only fought,
Nor Arts, nor Sciences by reas'ning sought.
    Walls, roads, and ruins, still attest the sway
They held o'er savagedom ('twas all they wrought),
    Fleeting such pow'r, and fast is its decay,
    But vic'tries over Nature ne'er can pass away.

Where now is Rome ? and where her vaunted might ?
    Where now the queenly mistress of the world ?
Oh baseless hold ;—she from the gilded height
    Of hollow royalty was headlong hurled

Into oblivion's sea, and o'er her curled
  Relentless waves, that wrecked the nation's name,
Whose aggrandising banners, oft unfurled,
  Waved but to elevate her murd'rous fame,
And left the savage worse than when midst them they
    came.

So with those glorious empires that essayed
  To flaunt their pow'r, from Time's assaults secure ;
For whose down-trodden millions tyrants made
  Debasing laws that served but to allure
To ruin's gulf, with silent march and sure.
  To Him, loud cries of vengeance rent the air,
He, swift as lightning did his wrath outpour,
  Then sank the monarch's heart in deep despair,
  As desolation reigned o'er all that once was fair.

How are the mighty fallen ! Tell, oh ye stones !
  Peace and goodwill ne'er touched the hearts of those
Who filled unworthily thy transient thrones,
  And rent mercilessly the world's repose,

Then smiled benignly on a realm of woes.

   Where are they all? forsooth! the stones declare

That nations die, and in their dying throes

   Some nobler paths for mankind they prepare,

     That elevate, and bonds of superstition tear.

A dying world! God from his lofty throne

   Beheld his planet rolling on thro' space,

Surrounded deep with sin-charged dark'ning zone

   That denser grew, as Time flew on apace.

Was there no light for the rebellious race?

   The Father's watchful eye with pity glowed,

Then council held, resolving to efface

   Mankind's illusions, He, from His abode,

     The darkness to remove, His only Son bestowed.

Expectant Judean hills echoed the cry,

   "That unto man The Wonderful is born,"

Nor nations joy, nor pompous pageantry,

   Welcomed the dawning of that blessèd morn

When sting of Death was of its terrors shorn.
　　Leap sin-clad earth! Light to mankind is giv'n,
And fettered sinners fling away with scorn
　　Satan's infernal gyves, that long have striv'n,
　　Ascendant to remain, and bar the way to Heav'n.

Thou blessèd yet accursèd land hast heard
　　A Calv'ry's cry, that on through time shall ride,
For thee, far better had His warnings stirred
　　Thy adamantine heart, than He, thy guide
Had been disowned and vilely crucified.
　　Not all disowned, for chosen pioneers
Heroically braved the pagan tide :
　　East, west, north, south, the infant faith appears,
　　Then fell the blindness of accumulated years.

As lightnings shoot athwart the murky sky,
　　Piercing the vault of Heav'n's o'ershrouding gloom,
So flew the burning words of God most high,
　　The wond'rous birth, the rising from the tomb,

The wond'rous love to say from final doom.
  A royal Roman heard the heav'nly tale,
His chiding conscience woke but to consume,—
  Believed he that it " almost" could prevail
  To shake his vain beliefs, and former sins bewail.

Then shook the base of all the pagan creeds,
  Then fell the idol o'er its glittering shrine.
Men heard, believed, and saw th' amazing deeds
  Of God in man—the new—the heart design.
No more to be unknown, but on will shine,—
  But yet the while by Satan's darts assailed,
That oft to death would holy men consign,
  Whose blood the seal of misery entailed
  Upon the mandate's source, o'er whom their faith
    prevailed.

What tho' the Crescent led in sensuous pride
  Its hosts of fanatics, that e'en would deign
To sweep the earth like an o'erwhelming tide,

And by the sword their bastard faith maintain ;
Their realm and vot'ries lie in darkness' chain.
   As rays of rising sun the mists dispel,
Before the Cross the Crescent sure shall wane,
     The weakened power and science' march foretel
     Their brighter intellects will soon their creed repel.

Far to the West the light increasing grew,
   And slow prepared the mind of man to rise
To higher aims, and his condition view,
   And Ignorance' creations to despise,
Seeking from Nature works of enterprise.
   Thro' envy's war, and 'mid each onslaught wild
That with subverted Rome inherent lies,
   Triumphantly it flamed, pure, undefiled,
   To priestcraft tyranny could ne'er be reconciled.

Thou loved Britannia, bulwarked by the sea,
   A Cæsar's host from conquered Gaul surveyed
With longing eyes, their prize erewhile to be

The Western Province by the Romans swayed :
Thy woaded savages, bold, undismayed,
  Waged war incessant 'gainst proconsuls skilled,
Who knew not that in thee deep underlaid,
  Slept instruments of pow'r that God had willed
  Would far outshine their own ;—now wakened ;—
    so fulfilled.

Thou favored isle, destined to hold thy place
  As first of earth, the mightiest and best,
Well thou receivedst the truth, and by its grace
  Did'st with an armour sure thy throne invest,
Irradiating aye the High behest,
  Thy generations rose to higher scale,
Panting to gain of Sciences the crest,—
  With liberated minds free to assail
  Obedient Nature's depths, and baser works curtail.

Through nations hates and wars, tradition bore,
  Attenuated tale of other days ;

A something strange, by Time unrusted o'er,
   Yet floating on, enveloped in a haze
Of centuries, and in its dormant phase
   Revealed how Hero, to mechanics prone,
Had proved that steam a wheel would turn and
      raise
   A weight ;—so, as the light dim dawning shone,
   'Twas Albion's sons revived this tale of years bygone.

'Twas noble Worcester's inventive mind,
   Who first as fact th' embryo pow'r did show,
With rude appliances ; and clear defined
   The boiler separate, from which did flow
The steam, then water from a depth below
   Was forced ;—none did the advent great proclaim
Of this, true birth of Superstition's foe,
   And forerunner of England's glorious fame,
   Who, blind with greatness now, unhonours
      Worcester's name.

He laid the base ;—again was pierced the veil,
   Thro' which the rays of Science faintly beamed.
Imbuing other minds who strove to scale
   The misty height o'er which the sunrise gleamed,—
When Savery appeared,—whose labours teemed
   With marked improvements on the primal birth,
Whose dawning influence he wisely deemed
   Would soon impermeate the darkened earth,
   O'erturning boorish labour by its cogent worth.

Onward and upward now ! more light the cry
   Of wak'ning men, who, with empiric skill,
Fashioned the fact, tho' ignorant of the why,
   And bearded Nature with determined will ;
Exultant day-dreams did with Hope instil
   A Newcomen's, then Papin's genius-mind,
Whose toiling reared the structure higher still,
   And left inventions, that, so well designed,
   Will through all ages be essential to mankind.

Slow spread the light to give presiding laws,
   And on sound theory perfection base.
Who shows Effect, yet graspeth not the Cause,
   A hazy and untutored mind displays;
Ambiguous and flimsy works they raise.
   From thee, loved Scotia, rose in splendour pure
The sun of Watt, whose genius all a-blaze,
   The mystery revealed, till then obscure,
   And crowned the edifice that ever will endure.

Scotland! 'mong all thy Gifted sons up reared
   In arms, in arts, and song,—a long array,
Outshining all,—none greater hath appeared
   Than wondrous Watt, who singly paved the way,
And brought the world under science' sway.
   What tho' a nation's laurels deck the brow,
To him the universe doth homage pay,
   Uplift thy honoured head, Old Scotland, now,
   He is thy greatest son, be proud for ever thou!

Behold the change ! see thro' the widened rift
  Britannia's greatness rise in plenitude ;
Now science' rays, disseminating swift,
  Dispel the darkness by their real good,
And generations rise with light imbued,
  Eager for other conquests, faint revealed :
Resistless in their march, all is subdued
  And made subservient to the pow'r they wield ;
  E'en Nature, paralysed, obediently doth yield.

Wide o'er the realm Wealth's speculations rise,
  Developing its treasures slumb'ring deep,
Our Hist'ry's builders, skilled in enterprise,
  Extend their schemes and other vict'ries reap ;
While smiling Science equal pace doth keep.
  Undaunted not, 'twas Bell the power applied
To brave the sea, and o'er the bosom sweep,
  Heedless alike of wind or adverse tide
  His little Comet sped—precursor of our pride.

A mighty leap was this ; no transient stride ;
  A great conception of still greater brain.
Old Ocean, frowning, saw strange monsters glide
  And flung its angry crests, that cleft in twain,
Fell impotent into its arms again.
  On ! on ! they go ; ye wildest winds be still !
They bid defiance to a raging main,
  Bearing the Light ; know ye they must instil,
  In minds debased, the truth, to all men, peace,
    goodwill.

The Earth, aghast ! beheld this victory,
  And, trembling, shook when Stephenson was born.
Humble of birth, he, in obscurity,
  Mated with poverty, and felt its scorn,
(True genius' children oft of wealth are shorn).
  A master mind was his, who did devise
Mankind's best boon, that like the sun in morn
  Banished the ignorance of statesmen wise,
  Who oft discern not good but with reluctant eyes.

Northumberland! thy laurels ne'er shall fade,
  His name and deeds peculiarly are thine,
Nations unborn will view the works he made,
  And venerate the genius of the Tyne,
Whose fame in Time's Valhalla aye will shine.
    See, whirling civilisers on the rail;
(Their mighty speed and banners white combine
  To vanquish bigotry, and rend its veil,)
  Compared with such a sight, all other triumphs pale.

Down with the rails! remotest climes surround,
  The earth encompass with our iron bands;
The whistle's scream will heathenism confound,
  Sole missionary best, for darkened lands,
And caste destroyer with no armed commands.
    Deep in the forest, 'mid their rude abodes.
The dusky natives view with clasped hands
  The harbinger of peace, with living loads,
  And deem the white man's pow'r far greater than
    their gods'.

Men such as these, have made our name sublime ;
　　Warriors of peace, their victories shall roll
Like an o'erwhelming avalanche thro' Time,
　　Increasing in its path, until the whole
Dark unbelieving Earth attains the goal
　　Of pure perfection ; then, with wild acclaim,
One cry alone resounds from pole to pole,
　　"That Light to all is given ;" men shall proclaim
　　His wondrous works and love, and bless His holy
　　　　name.

Forward, England ! to thee these sons were giv'n,
　　That thou mayst foremost be, and lead the van
In earth's redemption, as ordained by Heaven,
　　Spreading the word abroad to ev'ry man ;—
The word, that at Jerusalem began
　　To be first preached, for elevating those
Who fear to think, or their condition scan,
　　Who have no hope in life, no heart-repose ;
　　But blindly bask beneath Idolatry's dire woes.

On ! in the light, men higher yet shall soar,
  Wider extending their all peaceful views.
A Christian commerce busy on each shore,
  Still nations' fears and wars alarms subdues,
While each in love their kindred course pursues ;
  Raising their voices to the throne supreme,
One God, one Hope, each praying heart imbues,
  While angels from their realms with praises teem
  Of world Christianized, by aid of Mighty Steam.

DINNA GANG TO SEA

Gang na to sea the day, my dear,
    Ye needna stap the mast,
The waves beat high upon the pier,
    Wait till the gale blaws past,
Oh! let my love for thee prevail,
    An' mind oor bairnies three,
Sae lea' the boat, tak' doon the sail,
    An' dinna gang to sea.
        For lood is blawin' noo the gale,
          The white tap't waves I see,
        Oh! lea' your boat, tak' doon the sail,
          An' dinna gang to sea.

I dreamt last nicht, when in my sleep,
    A boat at sea upset.

I saw a face sink in the deep,

   His look I'll ne'er forget.

Sae gang na on the stormy main,—

   The feelin' maks me dree

That I will ne'er see you again,

   Gin ye should gang to sea.

      ' For lood is blawin', &c.

A doomed ship's drivin' in the bay,

   What should the pilots do ?

Launch oot their boat, an' sail away,

   An' try to save the crew,—

I couldna live were I to ken,

   That lives were lost by me :

Think o' the wives o' helpless men,

   Sae I will gang to sea.

      Tho' lood is blawin' noo the gale,

        An' white tap't waves ye see,

      We'll launch the boat, an' reef the sail,

        For I must gang to sea.

Mind not your idle dreams, my Kate,
   My bairnies weel I loe,
Yon ship drives on to awful fate,
   Oh ! think, lass, on her crew.
Sae dry your tears, I must away,
   Our boat will ride so free ;
When duty calls, I must obey,
   An' I maun gang to sea,
      Though lood is blawin', &c.

THE SHARK

WHY prowls the shark so long in our wake?
  Mark its ceaseless lightning sweep;
And why with fear doth the bravest quake,
  As it flieth through the deep?
For many a long day its course is steered,
  With ours o'er the trackless sea;
Far, far astern is its back-fin reared,
  See! see! 'tis now on our lee.
      The sailor thinks oft of the sea-swept dead,
          Of home, and those that weep;
      'Tis his only dread on lone watch or in bed,
          The tiger of the deep! the tiger of the deep!

He reigns supreme in his kingdom vast,
  The wide ocean is his home;

He scorns the waves and the Storm-King's blast,
   And he fearlessly doth roam.
An omen of death, he scents his prey,
   Weary he never can feel,
Haunting the ship by night and by day,
   He waits for his coming meal.
      The sailor thinks oft of the sea-swept dead, &c.

Perchance! 'tis decreed that death ere morn
   Will muster some gallant brave ;
And hammock-bound corpse in tears be borne
   By messmates to ocean grave.
A sailor's prayer, and a splash is heard
   Ascend on the morning breeze ;
No more astern is the back-fin reared,
   Of the despot of the seas.
      The sailor thinks oft of the sea-swept dead, &c.

SCOTLAND'S SANGS

Yᴇ needna' tell o' foreign lan's,
  Nor yet o' foreign bowers ;
Ye needna' chant your foreign sangs,
  For there are nane like ours.
Oor sangs, chairged wi' oor mountain air,
  An' bracing as the sea,
Eclipseth a' your foreign lays,
  An' aye will bear the gree.
      Raise high Scotland's banner of song,
        While on axis the warld aye turns ;
      Let Scotchmen aye toast when they drink,
        The sangs of auld Scotland and Burns.

Auld Scotland's sangs o' love an' war,
  Her mountains an' her men ;

Shall see a threed-bare warl' exist,
   An' will stan' foremost then.
Time canna' bring auld age to them,
   Nor will their beauties fade;
True Scotchmen yet unborn will sing
   The sangs their faithers made.
      Raise high, &c.

THE CROON

Sir C—— D——" I am a Republican."

SANDY *loquitur;* " Are ye, tho'?   Vera weel."

WE canna want oor croon, we winna want oor croon,

Republicans may rave an' rant, or try to blaw it doon.

There's men amang the heather roun' Victoria sune
  wad gather,

Wha ken fu weel it to uphaud, an' fegs, they wad ye
  lather.

  Sae we winna want oor croon, we canna want oor
    croon,

  Ye waste your win' ye silly fules, you're brayin' to
    the moon.

A month has no gane by, a' heard the nation's cry

O' sorrow for Victoria's son, when he was like to die ;

Death had an arrow pulled, his aim was sune annulled,
The Prince jinked clean a corner roun', an' Death
    was deftly Gulled.
    Sae he'll get his rightfu' croon, wi' a' your glaur
      an' gloom,
    Ye Brad-ilkites be wise in time, an' dinna fash
      your thoom.

Ye say it costs us dear, uphaudin' princely gear,
I wunner noo what ye subscribe—'tis unco sma' I fear.
Awa! to rural schules, ye'll see 'mang laddies rules,
That men wha preach without their brains are chief
    o' nature's fules.
    Sae we winna want oor croon, we sanna want oor
      croon,
    Ye waste your win', ye lunatics, you're brayin' to
      the moon.

Ye think ye muster strang, we'll lat ye rin alang,
Your tether will come to its en', wi' it we will ye hang ;

There's chiels ayont the Tweed that dinna fear to
bleed,

Sae Brad-awlites list to a freen', an' cease your daft-
like creed,

For we loe the auncient croon, oor auld historic
croon,

Gin forefathers did it maintain—wha'll be a
trait'rous loon.

An auld advice I'd gie—impossibles aye flee—

Or glaiket schemes that sure will lead to rank dis-
loyalty.

Mang cronies far aboon, in country or in toon,

Shines the man wha loes Old England, the Sovereign,
an' the Croon.

Sae we'll aye uphaud oor croon, oor ain oor
Christian Croon,

Rave on! Blaw on! Republicans, ye'll never ca'
it doon!

TRUE AFFECTION

Oh, Jeanie! worthy of my lays,
Whose eyes beam with affection's rays,
    And heart the same;
Thine be the honour, thine the praise,
Of happy home and lightsome days,
And cheering by thy winning ways,
    Your William.

The day I bless when ye were mine:
Darkness dispelled, new light did shine—
    I saw life's aim.
Thy love, based on affection's shrine,
New hopes begat, on grand design
Of love in life, without repine,
    That all should claim.

As time flies on with ceaseless wing,
We stronger to each other cling,
    In loving faith ;
Affection's hold still cherishing,
All selfish thoughts aye perishing,
The lamp of love aye nourishing,
    To end in death.

As ship rides out the fiercest gale,
Trusting to cable, not to sail,
    Devoid of dread,
So loving hearts can never quail
At storms in life, that may assail
Love's cable strong, no powers avail
    To break a thread.

FINCHALE

(A Fragment)

'TWAS Autumn time not long ago,
When freed from toiling business' flow,
Bent on seeking rural pleasure,
I was trudging at my leisure
Along the banks of bonnie Wear,
My lovely stream so sparkling clear;
Its Autumn song, so gaily singing,
Delight unto my heart was bringing.
From dell and glade the feathered throats
Vying sang out their evening notes,—
Dell and glade the path adorning,
Dell and glade for many a turning.
Ripened fields with sunset burnished,
Golden like, their treasure furnished;
Gently waving, brightly gleaming,

Ocean-like in slumber seeming.
My muse, exulting, rapturous swells,
At scene like this, (which few excels,)
Brings to flame the willing embers,
Leads me into homely numbers,
Gives my heart ecstatic flowing,
Wearside, the theme for fancy glowing.
Come! slumb'ring muse, awake and be
Trammelled with no vain rivalry,
Open thou thy mine of pleasure,
Let me rob thee of one treasure;
Bid me tell in accents simple,
Wondrous tale of ruined Finchale.

The sun far in the gleaming west,
With regal splendour sank to rest,
Hills and clouds with golden fringes,
Encircled were with dark'ning tinges;
O'er dell and glade the twilight falls,
From Cocken woods the night-bird calls,

To Hoo! To Hoo! a gruesome cry.
To Hoo! To Hoo! the woods reply.
Strange, solemn sounding, on my ear
Fell the low rumbling of the Wear ;
With stealthy strides th' approaching night,
Was mantling the last rays of light ;
Above, around, the dim pall hung ;
Surrounding woods deep shadows flung,
Sadly sighing, the evening breeze
Moaned weirdly 'mid the waving trees ;
Birds disturbed gave feverish twitting,
And from bush to bush were flitting ;
With loneliness my heart seemed flurried,
And wearily I onward hurried ;
My footstep's echo oft would start me,
As if some spirit meant to thwart me ;
I'd shivering stand and peer around,
Dreading to hear mysterious sound ;
Thus onward yet, tired, footsore, dull,
My mind with strange forebodings full ;—

Before me Finchale's tower appears,

The lonely link of bygone years :

Night-shrouded, weird, the ruins rose,

Meet haunt for dreaded spirit-foes ;

As veil of darkness ends the day,

So came dark thoughts with powerful sway,

Back, back to distant ages fled,

My living mind seemed with the dead,

Conjuring ancient pristine glory,

Abbots, monks, and cloistered story ;

'Mid wild imaginings enthralling,

Methought I heard soft footstep falling ;

As hedgehog coils its life to save,

My tongue contracted speechless clave

To drooping jaw, by fear unnerved,

While heart seemed from its region swerved ;

Keenly watching, keenly listening,

The dew of fear my temples glistening,

Straining eyeballs, hair all moving,

Of former sins myself reproving ;

As hills surround the stormy lake,
And 'gainst their base the wild waves break,
So lashed the mind its fear-bound zone,
With waves of thought still rushing on ;
I tried to run, from thence depart,—
My feet seemed of the earth a part ;
Fearing some sudden treach'rous doom,
I sidelong gazed on ruin's gloom ;
Oh cowardice so pure, refined,
Soon, soon you make men death-resigned ;
But lo! a footstep's stealthy tread,
Nearer approached from Finchale's shade ;
Some voice, borne on the sighing wind,
In tones encouraging and kind
Fell on my ear and loosed the spell ;
Returning thoughts vague fears dispel,
Returning thought new life instilled,
My heart with courage once more filled ;
"Stranger! what seek'st thou here I pray
By Finchale's ruins, hast lost thy way?"

" E'en so " I said, " thus pleasure's power
Leads oft astray at midnight hour,
By lovely Wear I longed to wander,
And view famed Finchale's ruined grandeur."
As sigh of winds from waving glade
Thou hear'st, but know'st not whence 'tis made,
As spirit-whispers heard in sleep,
That voice, from out the darkness deep,
So softly came, " Ay, Ay, wilt stay
With me until the break of day ?
Tired thou must be, and long thy fast,
Welcome to share my rude repast,
If Finchale thou would'st truly know
Come, Stranger, to my vault below :"
As mists from mountains downward roll,
Strange, humid feelings o'er me stole,
Longing, doubting, wishing, fearing.
A voice, but not a being appearing ;
'Tis not the age for magic spells,
No ghosts or spirits haunt our dells,

Long freed from dark oblivion's night,
We follow in celestial light;
Why should I fear, why hesitate?
Tho' weak in faith, tho' sinner great,
Come valorous thoughts with thee I'll brave,
Man, spirit, ghost, aye e'en the grave.
"Thanks, Unknown, thanks, I'll gladly share
Thy rude abode, thy humble fare,
Whoe'er thou be, where'er thou stand,
Show me the way, Come! take my hand,
I trust thee, though I cannot say,
What thou may'st prove ere dawns the day."
My words, scarce uttered, back seemed dashed
On sudden, shrieking, roaring blast,
So cold, so keen, so bound my breath,
My vitals crept with chill of death,
By subtle skill, by fell device,
I felt as if transformed to ice.
The frigid blast my temples kissed
While in my ear the words were hissed

" To thee Finchale I will divine."

Then unseen hands uplifted mine,

With adamantine grip were seized,

And all into a solid squeezed :

No flesh was there, no friendly feel,

My hands seemed held in bands of steel,

Ten bands of steel so hard, and dank,

At such a hold my courage sank ;

I vainly tried to speak, to plead,

My tongue essayed, but speech was dead,

O, Spirit blast, howl on and shriek,

I see, feel, hear, but cannot speak :

Fast o'er the sward by Unknown led,

Thro' murky ruins quick we sped,

Now up, now down, o'er mound and hollow,

Unwillingly I onward follow,

Fainting, swaying, stooping, flagging,

Unearthly power me onward dragging ;

Cold the blast still fiercely blowing,

Darkness seemed still darker growing,

Till slackening speed me silent told,
I'd some dread mystery behold ;
" Tired Stranger, come ! 'tis now we halt
Thy wish ; descend we to my vault."
One lingering look above was sent,
No ray of light in firmament,
No ray of Hope to guide or cheer,
No sound of speech to balance fear,
Within, without, above, around,
Cimmerian darkness reigned profound ;—
Groping, reeling, stooping low,
On rugged steps I downward go,
Cold was the cell, and putrefying,
Description true my pen defying,
Down, down, till on some broader stone,
I stood awaiting my Unknown ;
A gentle rustling by my side
Betokened presence of my guide,
A gentle breeze seemed o'er me blown,
I felt I did not stand alone ;

When suddenly appeared in sight
From farthest end a feeble light.
Dimly small, but growing clearer,
Far off, but quickly coming nearer,
Till filled the vault with sickly hue,
When horror! there stood in my view,
A being in dark mantle clouded,
Head, shoulders, all, completely shrouded,
Sleeves wide and long o'er breast were flung,
From tight drawn ends, there lifeless hung
Those fleshless hands, whose fierce embrace
Time ne'er will from mine own efface,
Ay! there they hung so ghastly yellow,
Surrounded by a greenish halo
Whose flickering upwards, unsustained,
Showed that some life in them remained,
(Not life of joy, of air, or earth,
But bloodless life of spirit-birth;)
Immovable, and still, it stood,
Till shining hands reached mantle's hood,

Each side those greenish lights displayed,

Rays cold and wan from them conveyed,

No mortal e'er saw such a sight,

And Finchale ne'er had such a night :—

" Stranger ! fear not," the Unknown said,

The promise to thee that I made

Will be fulfilled, and ye shall know

What Finchale was long time ago,

Ay ! I will pour into thine ears,

A wond'rous tale of bygone years,

Fear not ! tho' fear within thee burns,

Trust me—my proof—thy speech returns;

Tho' prisoned in these saintly walls,

Have faith ! whene'er thy courage falls,

What you shall hear, what you shall see,

Requires great courage, trust in me !"

'Tis done ! 'tis done ! swift through my brain

Warm thoughts dispelled fear's colder reign,

(Fear, craven fear, of sin the birth

You rob mankind of manhood's worth,)

Leaped high my heart with gratitude,
To this strange being, who silent stood,
Viscous, half-muttered words confused,
Recurrent came when tongue was loosed ;—
" Thanks, grateful thanks, to thee I owe,
I will trust thee, come weal, come woe."
The mantle's hood that dimly shone,
By glimm'ring hands was backward thrown,
Good Heavens ! there gleamed so green and dull,
With phosphorescent hues, a skull,
Vaguely grinning, gently nodding,
Of darkest deeds methought foreboding ;
No eyes were there, but fixed instead,
Two shining points, whose rays of red
Full straight and clear upon me fell,
And bound me all as with a spell.
Ay ! there upon those fleshless cheeks,
With wavy motions shone those streaks ;
Come Courage ! now I will thee muster,
To face this strange unearthly lustre,

From side to side he turned his head,

And hands on high he wide outspread.

Instantly sounds a solemn tinkle,

It is ! it is ! the bell of Finchale,

Then in a momentary pause,

I marked strange movement of those jaws,

'Twas smile-like, sorrowful and wan,

With sickly sigh, he thus began :—

  * * * * * * *

OOR AIN CLYDE

My memory brings ye back to me,
An' ca's up happy bygane days,
Whan by thy stream wi' youthfu' glee
We ran aboot thy broom-clad braes.
An' on thy banks in sunset's wane,
Fu' aft we've sang wi' glowin' pride ;
The hills echoed oor notes again,
Entwined wi' nicht-sang o' oor Clyde.
    Oor ain Clyde, oor bonnie Clyde.
      Your memory to me ever dear,
    Lang hae ye been oor ain pride,
      An' mony a dowie heart ye cheer.

Gie me the river rowin' on,
Whar aft in lover days I've strayed;
Twa hearts it saw in unison,

Alane it heard the love-vows made.
Aft hae I heard frae oot the fells,
The lavrock's love sang to its bride,
An' blackbirds' notes frae leafy dells,
Gae welcome to oor bonnie Clyde.

Oor ain Clyde, oor bonnie Clyde, &c.

The days o' youth I aft reca',
Whan roamin' far frae thy lov'd shore,
The thocht aye mak's the tear doon fa',
An' lea's me sadder than before.
Tho' bygane joys can ne'er return,
I'll loe thee aye whate'er betide ;
An' Scotchmen wha wi' hame love burn
Will aye sing o' oor bonnie Clyde.

Oor ain Clyde, oor bonnie Clyde, &c.

BOY MEN

Our chubby, laughing, rosy boy,
    Who knows no flight of Time ;
Is pleased with every newer toy
    And every jingling chime.
The tiny sparks or feathers light
    His mind with wonders fill ;
Great littles give his heart delight,
    And constant joys instil.

His little mind hath little thoughts ;
    His glee-shout, void of trouble,
Rings loud and clear as upward floats
    The shining soap-made bubble.
Up, up they beautifully rise,
    Some loftier flights will take ;

Pure pleasure fills his wondering eyes—
    Great sorrow when they break.

And thus like children men are oft
    More easily pleased than they ;
Aërial bauble borne aloft
    Will all their actions sway.
Inflated schemes bright points disclose,
    That temptingly adorn.
Men blindly hope to cull a rose,
    But ah ! they clutch a thorn.

Full oft child-men show childish power
    Of trifles o'er their minds.
That, captivating every hour,
    Life's worth and object binds.
The aimless manhood—boy designed—
    (Who all-wise vainly seem,)
May yet awake, but 'tis to find
    Their life has been a dream.

Life is not made of fancied joys,
   But stern realities ;
Its battles are not fought with toys
   Of flimsy qualities.
Perish boyhood's following ways;
   Strive to lead in battle van ;
Of noble aims the standard raise ;
   Upward—onward—be a man !

ADDRESS TO CONSUMPTION

Famed Nibbler ! to a vital part,
Maist cunnin' sharpener o' Death's dart,
Ye death in life ! wi' vampire art
    A haud ye tak',
Nae doctor chiel, wi' skill sae smart,
    Can drive ye back.

You're slow but sure, an' that's enough,
Oor wee bit breathie oot to snuff ;
An' tho' ane thinks their sel' gey tough,
    Ye are the ane,
Wha picks the flesh, wi' kittlin' cough,
    Frae aff the bane.

Ye gar the heart tak' mony a faintin',
An' aft the cheeks wi' roses paintin',

The hoast at nichts a' sleep preventin',
 Tho' unco sma'.
We think 'tis hardly worth lamentin'—
 'Twill wear awa'.

Aye, mony a body I've kent weel
Hae at your han's got gentle wheel
Intae some neebor warl' to feel
 Their promised due,
'Mang happy saunts, or else the deil
 An' his vile crew.

You're unco nice too, in your taste,
Young life-like fouk's gey aft your feast';
Wi' life ye tempt them to the last,
 Feent ony deein'.
They'll no believe they're gaun sae fast,
 Till aff gae fleein'.

An' no like fevered ends ava,

That aft the senses steal awa,
Syne in staps Death wi' unkent blaw—
    Ye canna tell't.
But thou ! Death's threeds ye spin sae sma,
    An' a' is felt.

Weel, efter a', it's nae great maitter
The way we throw aff oor vile natur'.
Tho' Death destroys the human creatur'
    Whan in the dust ;
The eftercome, wi' glorious featur',
    Should be oor trust.

THE GUID WIFE

WHAT maks a hame happy? what maks a hame cheery?
  What maks labour licht to the hard toilin' man?
Why, a couthie bit wifie that never is weary
  In keepin' things richt an' as snod as she can.
Her hoose is her hame, her guidman is her study,
  And tho' a' their income is aft unco sma',
Their bairnies aye cleadit are never seen duddie,
  Her shillin' gangs farther than ither fouk's twa.

Her smile in the mornin' the guidman aye cheerin',
  As blithe wi' the daylicht he gangs to his toil,
Stan's ever before him, an' smooths whan appearin',
  Asperities dreich in the warl's turmoil.
"Aye up wi' the birdies, whan sun's rays are peepin'."
  Has been lang wi' her a weel-acted-on rule;

F

The house wark is done when her neebors are sleepin',
 The bairnies, syne parritched, are dressed for the
  schule.

Frae mornin' till e'enin' she ne'er is fand idle,
 But darnin' at stockins or mendin' o' claes ;
A queen amang neebors, her tongue needs nae bridle,
 Her guid sense is proof to their gossipin' ways.
Her life is nae burden, time deals wi' her lichtly,
 An' keeps her aye free frae a' harassin' cares ;
Tho' clad in the hamespun she aye looks fu' sprichtly,
 Contentment is boun' wi' the smile that she wears.

At nicht whan the faither is hameward returnin',
 The bairnies a' meet him a wee on the road,
He hears in the daffin', frae hearts that are burnin',
 A welcome o' love to his theekit abode.
An envy he is to his far richer neighbour,
 Whase life ne'er was brichtened wi' love's faintest
  streak.

Wha looks wi' disdain on love's welcome to labour,
  He kens na that love can be pure under theek.

The guid wife wi' pleasure marks a' their caressin',
  Syne lovingly tells them to set themselves doon,
Tho' humble their supper, they aye ask a blessin',
  An' thankfu' gie praise to the Giver aboon.
The supper bein' endit, they sit roun' the ingle,
  Some learnin' their lessons, some playin' wi' toys,
Sae cosy an' happy, nae thrawness can mingle,
  Whar mitherly fondness rules a' their wee ploys.

Nae vile gowden notions their daily lives cloudin'
  They kenna the weight o' wealth's sorrow or care ;
Pure love aye is free frae ambition's o'er-shroudin',
  The little they hae is ne'er curst wi' "some mair."
Tho' dwallin's be stately, tho' gowd be a plaything,
  Tho' fouk seek in riches to mak a great name,
The tapetless silken-clad dame is a naething
  Compared wi' the wifie whase love lichts a hame.

SPRING

GRIM Winter caught Spring in his cauld, icy han's,
  An' fain wad hae smored her whan born,
He roun' her wee thrapple had boun' snawy ban's,
  The warslin' wee thing gat them torn.
Tho' scanty o' happin', an blae wi' the cauld,
  'Maist nakit in mony a storm,
She lauchin' gaed loupin' roun' Winter sae bauld,
  An' kicked him to mak hersel' warm.
    She lauched an' she loupit, she ran here an' there,
      An' shook a' his snaw aff her pow.
    Syne as she gat stronger, kept muttrin' sae clear,
      " Fy on ye, auld man! I maun grow!"

The wee wean grew aulder, an' aften wi' joy
  Wad row a green gown roun' her waist,

The blust'rin' auld carlie, wad sune it destroy
    An' deck the puir thing like a ghaist,
Her feetie war frostit, sair, sair, did she greet,
    An' shiverin' an' toddlin' her lane,
Ae day,—frae aboon—some sma' freens she did meet
    Wha pitied this young raggit wean,
        She lauched and she loupit, she ran here an'
          there, &c.

She tell't them her story aboot the auld man,
    An' hoo aft to kill her he'd try,
The wee freens a' leuch—an' crapt into her han'
    An' whispered, "Ye'll noo gar him fly."
He aft in his anger threw snaw doon upon her
    She cared na' but shook her wee neive,
The han' fu' o' sunbeams gar'd snaw tak a scunner,
    Sae then the auld man took his leave.
        She lauched and she loupit, she ran here an'
          there, &c.

Tho' sair was the han'lin', she thrave an' look'd braw,
   An' glamoured a' things wi' her een,
'Twas no like her beauty, bein' clad aye in snaw,
   She aye lookit best in the green.
She wove her a mantle o' bonniest hue,
   An' donned it wi' smiles unco coy.
Wharever she wandered sweet flowers soon grew,
   An' wee buds soon keekit for joy.
    She lauched and she loupit, she gaed up an' doon,
      An' birdies to her aye did sing,
    O' green leaves an' flow'rets she made her a croon,
      A' welcomed the braw lassie, Spring.

LAND OF THE MOUNTAIN, FLOOD, AND BRAVE MEN

LAND of the mountain ! once more I behold thee !
My heart is exulting as o'er thee I roam,
Recalling the bygones that fondly enfold me—
Still dear are the memories of childhood's bright
home.
Who would not love thee, thy glories oft telling ?
Or who would not gladly claim thee as their own ?
Beats there a heart with no proud feelings swelling,
If treading thy mountains majestic and lone ?
Scotland, loved Scotland ! my country, my
pride !
The home of the happy, the free, and the
brave !

Mountains, loved mountains ! where tempests
   preside—   .
Emblems of freedom ! Oh, there be my grave.

Land of the wild flood ! your beauties I cherish ;
  Thy cataracts leap with delight through thy dells,
Their mountain-made music gives songs, that ne'er
    perish,
  The strains of whose wild notes weave round thee
    heart-spells.
Loved are thy wild floods, and dear are thy moun-
    tains,
  So grandly and proudly they rear their blue crests ;
Down from their dark sides dash wildly thy fountains,
  Awe filling the vision—love filling our breasts.
    Scotland, loved Scotland ! my country, my
      pride ! &c.

Land of the brave men ! thy valorous fathers
  Oft victory boasted on many red field ;

Bright be those laurels whose beauty still gathers

Sons valiant around thee, who never can yield.

Land of the mountain, of flood, and of brave men !

Long, long shall thy sons your famed prestige
maintain ;

When Scotsmen lack virtues no heather shall wave
then—

Farewell to auld Scotland—her glories shall wane !

Scotland, loved Scotland ! my country, my
pride ! &c.

A MIDNIGHT SCENE IN SUNDERLAND PARK

" My Son, aye keep awa frae whisky,
An' 'twill keep you frae mony a pliskie."

NEW SONG.

OH! had I but the Muse's spell,
In polished language I could tell,
O' strange! strange scene that me befel
        The ither nicht,
As midnight struck on toon-clock bell,
        Increasin' fricht.

When freen's forgather, an' late sittlin',
An' aiblins are themsels forgettin',
Wi' drink gaun in, an' sense oot-lettin',
        An' robbin' truth,
They think na on the morn's regrettin',
        Or geisind mouth.

I had been drinkin' at the "Queen's,"
Wi' twa three ither jolly freens ;
The time flew by wi' unclipt wings,
   Owre quick the speed,
The hame-ca' ilka clock mainteens,
  To save oor heid.

The shak-han' guidnichts soon war owre,
Each ane set aff for his ain door,
An' for ae stap I aft took four ;
   I wis my 'lane,
While toddy-sweat began to pour
   Owre back an' wame.

The scant licht frae a deein' moon,
Was loth to leave the warld sae soon,
Reflectin' shadows a' aroun'.
   In color grey,
An' gripp't on housetaps up aboon
   Wi' better stay.

Methocht I through the Park wad gae,
To clip a bit aff langer way,
An' bring me hame wi' shorter stay,
                To my ain Jean.
Or cool my head frae toddy's sway,
                An' clear my een.

Oh ! Sunderland, proud may ye be
O thy braw Park—nane e'er will see
A lovelier—else they tell a lee
                O hue sae dark ;
I've seen a few, but nane like thee,
                Thou bonnie Park.

By Havelock's monument I strayed,
An' sat me doon in his loved shade,
Reflectin' on this " great unpaid,"
                A red-tape plaything,
Wha battles won, an' kingdoms made,
                An' a' for naething.

My whiskied fancy was beginnin'
To ca' up red coats, wha war thinnin'
The lang black croods o' Sepoys, grinnin'
    Wi awfu' glee,
But Havelock wis the battle winnin',
     The blaiks did flee.

The fancied battle's stirrin' soun'
To end cam quick—for frae the grun'
I heard terrific roars aroun',
     Infernal rise,
An bizzin' tremor soon begun
     My heart to seize.

I listened, an' frae oot the dell,
(Whar lovers coort, an' secrets tell)
Loud shrieks arose—then quickly fell,
    An' dee'd awa.
Like worm I crawl'd owre to the rail,
    An' sicht I saw !

A MIDNIGHT SCENE

There in the dell, beneath the hicht,
A bleeze shone roun' wi radiance bricht,
An' threw a glare to left an' richt,
        As plain as day ;
I thocht foul wark's gaun on the nicht,
        Or ither play.

In centre, on a flame-bound chair,
Wi serpents wrigglin' in his hair,
Ane sat fu' fiery blue an' bare—
        It was the Deil !
I saw a lang tail in the air,
        An awfu' chiel.

A kingly wave o' han' he made,
Instanter rose up frae the glade,
Fu' fifty deils, wha him obeyed,
        Wi magic pass ;
Anither wave, an' they a' gaed
        Doon on the grass.

While in the air vile creatures flew,
Some red, some green, an' streaked wi blue,
They a' had tails o' fiery hue,
       An' ugly een,
An' flames cam' oot o' ilka mou',
        Wi' dartin' gleam.

At door, a wee owre to the richt,
Twa deevils stood—ten feet in hicht,
An' shook their tails wi' great delicht
       At yells inside,
Shrieked by some prisoners catched the nicht,
        An' firmly tied.

A deevil jury were empannelled,
Wha yelled as if they were unkennel'd ;
Their debtors will be sairly haunel'd
       I shiverin' thocht,
They'll be devoured alive, or mangl'd,
        If oot they're brocht.

" Bring oot the first ! the Chairman cries,
" Lang, lang the rascal's lived on lies,
An' years ago we took his size,
                    To us belangin',
To him it will be nae surprise,
                    For a' his wrangin."

The twa ten-feet anes gae a skirl,
An' ope'd the door wi' thunderin' dirl,
Syne brocht ane oot wi' dizzy whirl,
                    'Mid smoke an' flame ;
Puir chiel ! his mou gae mony a twirl.
                    I kent his name.

Auld Satan gae a kingly smile,
As prisoner stood between the file,
In voice ye wad hae heard a mile,
                    He thus began—
" Arraign'd at last ! ye vilest vile
                    Disgrace to man.

" Stand up ! thou ignorant of the laws,

Ye've aye shone in a fallin' cause,

A name ye hae,—for windy blaws,

   Ye empty blether.

It's a' prood-flesh came in your claws,

   Ye lawyer feather.

" Know ye ! I lo'e ye as a brither,

An' fain wad save ye—for oor mither,

But ends o' justice I ne'er smother,

   As ye hae dune ;

Relations always give me bother,

   An' ye are one.

" Say, imps : What sentence will we levy ?—

Lawyers can stand it unco' heavy—

I vote—consign him to the bevy,

   At warmest portal.

Agreed ! there's none will ever save ye,

   Ye soulless mortal.

G

" Advance, an' put on him our crest,
So, when enjoyin' his het, het rest,
My freens, ye'll treat him to the best
                    We hae to gie,
An' strive ye hard him to molest—
                    Obligin' me."

In bucket lairge, wi' hue o' pink,
They toom'd in bluid, for markin' ink,
Brocht pen wi' mony a twistit link,
                    Syne on him seized ;
Whan, lo ! he looked like made o' zinc,
                    Or galvaneesed.

A dip ! ae stroke !—a fearfu' yell !
An' deils an' goblins on him fell,
His last words I will never tell,
                    Or deein' look.
Thocht I, that's ane ta'en straucht to h—l,
                    An' brocht to book.

Owre to the door he quick was ta'en,

Four han's appeared wha dragged him in ;

His deein' yells aboon the din,

      Terrific rose,

While deils aroun' gae mony a grin,—

      The door did close.

I thocht this Lawyer aye was bold,

Judgin' frae his Police Coort scold ;

He look'd the nicht, abject an' cold,

      He noo is richt ;

Lawyers are made o' saft, saft mould,

      I saw the nicht.

Auld Satan shook his towsie hair,

His lang tail waggin' roun' his chair,

An' crossed his arms wi' kingly air,

      An' glowin' pride.

I plainly saw there was some mair,

      Gaun to be tried.

"Bring oot the next!" was roared again ;
The ten feet warders, unco fain,
Anither brocht, wha yelled amain—
    I kent him weel,
To mony he's gien nichts o' pain,
    A Doctor chiel.

Their wings owrehead the goblins flapped,
As Satan spak', an' his han's clapped,
"Ha! Ha! sometime syne we had mapped
    End o' y'er race,
The roots o' life fu' aft ye've sapped,
    Wi' smilin' face.

"Ye've poisoned thousands wi' your potions,
An' mony wi' mercurial lotions,
Ye dosed the public wi' your notions,
    An' mony wranged ;
Sins o' herculean proportions
    To ye belanged.

" On mony a mither's tears ye've smiled,
Whan lookin' on her deein' child,
The deed y'er ain—ye've aft beguiled
   A mither's heart ;
Life-givin' lees, by Doctors' tell'd,
    Fause hopes impart.

" Come, jurymen, you must not swither,
This unhanged 's run oot his lang tether.
Our love o' duty ne'er can wither,
   Or weaker grow,
Doctors and Lawyers burn thegither
    Wi' bonnie lowe ! "

The deevil jury gae assent,
Deils yelled, an' goblins owre him bent,
Auld Satan gae a sidling sklent,
   An' tail was mudgin'.
Owre 's face I thocht some pity went
    For th' poor surgeon.

Oh! what infernal din arose,
As roun' him quick the Deevils close;
My fegs! he'll noo get such a dose
        He ne'er did taste;
Auld Satan's drugs gie warm repose,
        An' lang they last.

Their crest they quick put on his skin,
Decomposition soon set in,
The color blue, the froth ower chin,
        An' chatterin' teeth!
I'll ne'er forget deil's doctorin'
        Till my last breath.

Wi' speed o' lichtenin' to the door,
The ten-feet warders whisk'd him o'er,
It opened wide, an' smoke did pour
        In volumes black.
The Doctor chield I saw no more—
        He'll ne'er come back.

Auld Satan, wi' his tail for hammer,
Thrice struck his chair—the fiendish clamor
Was hushed by a', like breeze in summer,
   'Mang summer trees ;
The flames increased, sae did my tremor
   O'erpow'ring seize.

I saw him sit in regal pride,
Twa raws o' deils on ilka side,
An' mony imps, wha keenly vied
   To dae his wull,
While goblins sailed in circles wide,
   Owerhead the haill.

Again he spak : " bring oot the pair
We hae reserved oor joys to share,
We'll let them live, but they must bear
   Oor weel kent crest ;
They'll ne'er forget wha's in the chair,
   While life doth last."

The goblins flew—now up, now doon—
The deevils yelled, as dancin' roun',
An' Satan leuch at bonnie soun',
                    An' whisked his tail.
The imps o' deevildom like fun,
                    As weel's mysel'.

The warders shook their tails apace,
An' pleasure shone on ilka face,
Frae oot the door, wi' deil-like grace,
                    An' freenly air,
O' shiverin' mortals brocht a brace,
                    A weel-matched pair.

" Come! Editor an' Parson meek,
Stand up! 'tis not your lives we seek.
My work ye do, frae week to week,
                    Unkent to you ;
Your actions o' mysel' aye reek,
                    An' gie me due.

' The heart-blight o'er the land broad-cast
Ye both sow well—my trade 's maist passed
Into your han's—long may it last,
                Ye deevil twins,
No opposition I ere cast
                In way o' sins.

" To show our sense o' your great worth,
We'll mark you both, an' send you forth
Once more, to do upon the earth
                My dark designs ;
Your language, mine—you'll ne'er have dearth
                O' deevil's signs.

" My imps, be quick ! 'tis nearly one !
Bring out the brand—it must be done.
This Editor an' meek Parson
                Our brand must bear !
Editors an' Parsons are our own
                In loving care."

Surroundit' by the deevil bands,
Their heads held back wi' deevil hands,
While goblins screech'd at chief's commands,
   I then did see,
A warder wi' the brand o' brands,
   A red-het D.

In Satan's hands the brand appeared,
In twinklin' he their breasts had seared,
The devils' roun' but faintly cheered,
   Wi' smothered yell ;
Satanic brandin', I had feared,
   The pair wad kill.

Auld Satan flung the brand in air,
An' cries, "Away ye deil-marked pair,
My pleasure do, an' ye will fare
   A hantle better ;
An' do ye all that I would dare
   Men's minds to fetter."

They stood, released, tho' sickly white,
An' edged awa, wi' shaky fright.
A wee atour, wi' a' their might
        Began to rin,
An' efter them went deils good night,
        Wi' mony a grin.

The deils an' imps danced roun' their king,
An' goblins flew wi' rapid wing.
A deil-made sang they a' did sing,
        Wi' voices hoarse ;
Auld Satan's sangs an' dances bring
        A feelin' coarse.

A' shiverin' wi' the deil-bound spell,
Frae oot my grip a big stane fell,
Wi' mony a thud, into the dell,
        Amang them a'.
Imps shriek'd, deils roar'd, an' Satan's yell
        Was warst ava.

I scarce had time to tak a thocht,
When roun' my head the legion focht,
I fancied 'twas my brains they socht,
　　　　　To lay on sward ;
My life, they'll see, will be dear bocht,
　　　　　I'll dee gey hard.

They tugged, they pu'd, they nip't, they knocked,
Ae strong ane nearly had me choked,
I tried to shout,—they only mocked
　　　　　My twistin' mou' ;
My han's an' legs securely locked,
　　　　　What could I do ?

Wi' pleasure great this strong deil spak,
As he kept rivin' at my back ;
The voice !—my ain !—made me a' shak,
　　　　　An' reason brocht ;
This deevil spak in mither talk,
　　　　　Oh, cheerin' thocht !

"Get up ye fule!" I kent he said,
(His big han's liftin' up my head,)
"Get up! an' quick be off to bed."
   This wis a hobby.
I looked an' stared! wi' mou ootspread—
   It was a Bobby!

My tongue, tho' stickin' to my mouth,
Wi' eftercome o' whisky's drouth,
Was loosed,—says I : "Hae ye, forsooth,
   Seen the auld Deevil?"
"Be off!" I got, in voice uncouth,
   An' sae unceevil.

Wi' fear I tottered down the hill,
An' cuist ae look into the dell,
But a' wis dark—sulphureous smell
   Hung i' th' air.
The deil aye hides his wark fu' well,
   Wi' cunnin' rare.

The midnicht win' felt unco blae,
As trachlin' hame my lanely way,
Resolved I nae mair oot wad stay,
        Till clock struck one ;
An' frae Scotch whisky keep away,
        Like decent man.

A RETROSPECT

As days and months then longer years,
Roll o'er me one by one,
I ask myself with gravest fears,
Have I my duty done?
What have I done? Have I fulfilled
Man's highest destiny?
Have I thro' life sin's passions stilled
By deeds of purity?
Ah no, it is not thus with me,
I fear, I know, that deep within,
My heart enveloped is in sin.

What have I done? Why I have wrought
In labour's sternest field,
My sweat-stamped pittance, dearly bought,

Would sin-stamped pleasure yield.
Of years I barely count two score,
Yet backward as I gaze,
I see a life deep-furrowed o'er
With sin-recording ways,
And cannot point with pride or praise
To one bright spot that would appear
As oasis in a desert drear.

What have I done? Why I have borne
Adversities full keen ;
Desponding not, I found some morn,
Would dawn with cheering sheen.
Men and their ways I never feared :
With sail spread to life's wind,
I madly o'er its sea careered,
Yea, seldom looked behind,   .
My eyes to all but sin were blind :
Yet good instilled at mother's knee,
Oft rose before me chidingly

What have I done ? Why I have had
My share of ups and downs,
Met friends a few,—but many bad,
Whose smiles but glossed their frowns ;
Friendship ! how false, with me, 'most dead,
Oh, base it is around,
'Tis mighty Self that reigns instead,
Heart-virtues are not found ;
They cannot root on selfish ground ;
Few hearts can harbour Virtue's call
For man is Self, and Gold is all.

What have I done ? Why I have fled
Thro' life without an aim.
No good ambition e'er had shed
One ray that I might claim ;
Rebuking shadows of the past
Tell me of misspent years,
My might-have-been, is upward cast,
'Mid deep regret and tears ;

The hidden future dark appears,
While silent whisp'rer ever saith—
Think, think, on what comes after death.

These have I done, hush ! now I've reared
Within my breast a shrine,
For one who savingly appeared,
A being, pure, divine.
Then vanished darkness from my mind
I breathed another life,
Then reason's sway, so calm, refined,
Allayed my inward strife,
The present is with beauties rife ;
When purer thoughts this life began,
Strange ! up arose my other man.

Methinks good angels conclave held,
Where Pity, pitying sate ;
Far from their regions they beheld
A waif of self-made fate ;

Unloved, uncared, forlorn, alone,

This groping mortal sighs,

Till messenger from Pity's throne

Assumes the human guise,

Bearing to him her kindred ties ;

Then flung he lower aims aside,

Then found a comforter and guide.

Yes ! Love dispelled my long, long night,

(Nought can its power impede)

Now is my goal, to live aright,

In thought, in word, and deed.

Back rush my thoughts, oh what a void

One ray from her displayed !

A life nigh wasted and destroyed

Is vividly pourtrayed,

Primal Resolve rose from the shade ;

Then did its infant form proclaim,

Some good may yet adorn my name.

THE HARRIED NEST

OH! who with unrelenting breast,
Has robbed thee of thy young,
And left this little cosie nest,
An emblem of their wrong?
See! see! the loving parent pair,
Sit desolate and lone,
Hear anguished chirrups in the air,
'Tis their bereavement's song.

And mark from overhanging tree,
Short, feverish flights are ta'en,
The parent calls, unanswered be,
No loved will come again.
Vile ruthless hands who'd rob a nest
And desolation bring,
The brutal feeling once possessed,
Thro' life may to them cling.

ROKER CLIFFS

To Roker Cliffs let us away,
While yet the shining sun
Reflects so bright the cliffs' proud height,
On sparkling waves that run.
Along the silv'ry sand-bound beach
We'll spend the summer day,
And we will bathe in curling wave
By Roker Cliffs so grey.

By Roker Cliffs I oft have strayed
And viewed the driving storm,
The frowning rocks, the great wave mocks,
And breaks its sweeping form ;
The Cliffs so grim the white sea-foam
With a feathery down bedews.

High, high in the air, 'mid tempest's glare
Sweep round the wild sea mews.

O Roker Cliffs, with awe I view
Thy pillar'd caverns grand,
And wonder oft, as I gaze aloft,
If reared by Titan's hand.
The sea may lash thee with its might
And Time thy face may tear,
But sea and Time, tho' both combine,
Thy beauties cannot wear.

A SCENE IN THE HIGH STREET

Time—Saturday Evening

DEEP-BEARDED, helmeted, and dressed in blue,
Of such a form that he might well be proud,
Unblushingly, before my pitying view,
I saw him drag on thro' the busy crowd
A half-drunk woman, whose entreaties loud
Comingled with the cries of one who clung
With childish fondness to the parent's side ;
Relentless he, no pity could be wrung
From such a giant, who with glowing pride
Smiled on a mother's and a daughter's tears
With look triumphant, as he them did trail.
Unmanly sight was this ; e'en in my ears
Still rings that child's heart-breaking plaintive wail,
"Oh dinna tak my mither to the jail!"

MY PILOT

LONG o'er the sea of Life my course unsteered,
Chartless and rudderless, with ev'ry gale
I bounded on, no haven of rest appeared ;
The canvas rent, the timbers getting frail,
Ah! I had need of help from friendly sail ;
E'en sunlit days seemed dark, all, all was gloom,
No guiding star was mine, no cheering ray
To show life's breakers, or avert the doom
Of courseless bark upon an aimless way :
When from afar, by guardian angels sent,
Heading for me, a Hope gilt craft is seen.
I shout, Save me ! its pilot gives assent,
Light dawns again ; life's waves have golden sheen,
I reach loved port, being piloted by Jean.

GREE BAIRNIES, GREE

HECH me but I'm auld noo, wi' warslin' thro' life,
   An' bare shaks my pow to the win's o' three-score ;
I canna be fasht noo wi' bustle an' strife,
   But pechin', contentit, sit by my ain door.
Tho' auld, auld in years, fegs my mind is aft young,
   An' bairnies a' roon me I aye like to see,
Their daffin' remin's me whan my grannie sung,
      "Be guid noo my bairnies a' gree, aye a' gree.
         Whaurever ye gang, or whatever ye be,
         Be sure wi' your neebours, my bairnies to gree.

Ou ay ! whan recallin' my ain youthfu' years,
   I ponder an' wish I could live them again,
Hoo wise wad I be, wi' nae heart-braks or fears,
   That sel'-made hae boun' me, sin' I was a wean ;

Hoots! hoots, it's nae use to be wishin' this way,
   Ye ken I am haverin', for that canna be ;
Aft blin' is life's mornin', till closin' o' day,
   Whan hoastin' an' helpless we're glad then to gree.
     Whaurever ye gang, or whatever ye be,
     Be. sure wi' a' bodie, my bairnies to gree.

Weel, weel dae I ken that life's faught aye is warst,
   Whan fause independence fouk tries to mainteen,
Nane cares for a " wad be " wi' michty sel' curst,
   Wha looks aye wi' doot on the smile o' a freen.
A' honest endeavours are aye weel assistit,
   Whan clink o' the heart can be read in the ee' ;
Life's drag is unscrewed whan a' passion's resistit,
   An' whan wi' oor fellows we gree, aye a' gree.
     Whaurever ye gang, or whatever ye be,
     Be sure guid attends ye, whan wi' a' ye gree.

Waes me! as I look at mysel', hoo I think
   On life, an' its lessons, an' a' I've gane thro',

My toilin' an' fechtin', scarce brocht me a blink,
  O' calm heart-repose, or o' happiness true,
Yet I'm no alane, for maist fouk, rich an' puir,
  Hae seldom the broo frae some sma' wrinkle free ;
They wadna deep furrow gin a' wad be sure
  To be aye as bairnies an' gree, aye a' gree.
    Whaurever ye gang, or whatever ye be,
    The wrinkles lie lichtly, whan wi' a' ye gree.

Tho' auld be the warl', 'tis strange hoo the sense,
  O' them wha live on it is still unco sma',
Mankind coortin' follies at mankind's expense,
  Increaseth the mis'ry that hings 'roon us a'.
Hoo happy we'd be, baith whan young an' whan auld,
  Gin nations an' fouk wad a' selfishness flee ;
Upliftin' o' hearts gie's a blessin' untauld,
  An' efter gude waits them wha live aye to gree.
    Whaurever ye gang, or whatever ye be,
    Forget na, that Death vera sune maks ye gree.

JEANIE YE KEN

OR, AFF WI' YE SETTIN' SUN

BROON Autumn had cheenged into winter's grey tintin',
   An' bare waved the trees to the cauld evenin' blast;
The red sun thro' snaw cloods was timidly glintin',
   An' cuist its last ray ere it sank in the wast.
Fu' lichtsome I hied me awa to the shielin',
   That nestles sae snugly far doon in yon glen;
Nae gloamin' or cauld nichts could dauntin' the feelin',
   My heart had lang cherished for Jeanie ye ken.

    Aff wi' ye settin' sun hide in your cloody nest,
      Nocht dae I care tho' ye lea' me in gloom;
    I ken a brichter enshrined in a lovin' breast,
      Rays sae uncheengin' my path will illume.

Fast on thro' the heather the lane moor adornin',
   An' doon by the howe whaur the fairies aince danced

My heart tho' lood beatin' was craven fear scornin',

    Tho' sometimes aroun' me I warily glanced.

Love's wings bore me onward, the future divinin',

    I thocht mysel' happiest an' blest o' a' men ;

As far thro' the gloamin' I spied a licht shinin',

    'Twas placed in the winnock by Jeanie ye ken.

        Aff wi' ye settin' sun, hide in your cloody nest, &c.

Alane by the gairden yett, longin' an' listenin',

    Her heart unco fu' o' love's fondest alarms.

Stan's Jean, in whase black ee, hope's bricht tear is

        glist'nin',

    That sune is dispelled as she fa's in my arms.

Fa' deeper nicht darkness! blaw caulder nicht breezes!

    I bauldly defy noo the feelin's ye sen' :

Nocht pu's at the heart strings or mair pleasure gie's us,

    Than wooin' a lassie like Jeanie ye ken.

        Aff wi' ye settin' sun, hide in your cloody nest, &c.

Her head on my shouther sae gently reclinin',

    An' upliftit een,—tuts!—wad melt a whinstane,

The grip o' her saft han's that mine were entwinin',
   Proclaimed that the lassie's sel' was a' my ain.
Oor vows, an' oor whisp'rins o' love for each ither.
   A feast o' pure love joys ; sune cam' to an en',
For oot frae the door cam' the voice o' her mither,
   "It's time ye were beddit noo, Jeanie ye ken."
     Aff wi' ye settin' sun, hide in your cloody nest, &c.

The ca' o' the bodie oor pairtin' recordin',
   Fu' closer we clung, tho' in sorrow gey sair,
To end a' thae grievins an' joy be affordin',
   I thocht it was best to be pairtit nae mair.
Sae say the word dearest ; tho' humble my leevin',
   Ye ken love can dwall in a but an' a ben,
I'm sure your consent will be blessèd by heaven ;
   She whispered "Weel, weel," did my Jeanie ye ken,
     Aff wi' ye settin' sun, hide in your cloody nest, &c.

"Gude nicht to ye Jeanie, I'll hameward be gau'in',
   Ae kiss as a pledge that for ever you're mine,

I heed na the darkness, or cauld win's a-blawin',
  The licht o' your promise will roun' me aye shine.
The glen may be eerie, an' weet be the heather,
  An' lood be the scream o' the startled moorhen,
I carena, for ken ye, neist week we'll foregather,
  An' married I'll be, to my Jeanie ye ken."
    Come then ye risin' sun, break thro' your cloody
      nest,
    Shine ye in brichtness an' smilin' rays len' ;
    Walcome the mornin' o' day to me ever blest,
    I tak hame the lassie ca'd Jeanie ye ken.

FLOWERS

In the walks of Poesy,
   Such a garden rare,
Culling flowers rosy,
   Ever bright and fair,
Flowers ever bonnie,
   Flowers ever sweet,
Flowers full of honey,
   Poet's garlands meet.

Say! is aught else so pure
   As the dewdrop's sheen?
Say! can aught ee'r endure
   Winter's cold blasts keen?
Yes! purer far the flowers,
   Where love's dewdrops cling,
Where winter's withering powers
   Give them nourishing.

Flowers ever growing,
   Ye are so divine,
Ye are worth the knowing,
   Oh! do ye be mine;
I'd be always with ye,
   Flowers loved so well,
Be ye ever for me,
   Lasting immortelle.

THE WAYS O' WEALTH *

A' rich-blin' hearts will ae day rue,
The unchristian courses they pursue,
Their worshipped wealth will prove too true
      To be Hell's snares ;
A lang, lang time till Heaven be fu',
      Wi' souls like theirs.
Modern Mammon.

Dear Beattie, loved an' best o' men,

To wile an 'oor I'll tak my pen,

An' in my screed I'll lat ye ken

      My rhymin' reason.

My Muse is loth her powers to sen'

      This winter's season.

I fear I hae her will been scornin',

An' frae me noo her face is turnin'

Gin it be sae, I'll gang in mournin',

      Or penance dae,

* Vide National Debts, *Macmillan's Magazine*, January, 1872.

Her smile—neist Jean's, my life adornin'—
      I aye maun hae.

Wi' heart repentance at its hicht,
I vowed, and prayed to her last nicht,
Aince mair to lat me drink delicht
      Frae out her horn.
I couldna sleep, till dawned the licht
      O' early morn.

Weel, efter a', the cutty's kind,
An' tae her whims I should be blind,
I woke,—and fand fears thrown behind,
      For ever hidden,
Oot cam my thochts, frae lowin' mind,
      At her ain biddin'.

Carte-blanche I got, she bade me revel,
'Mang warldly things baith gude an' evil,
An' note the grips o' mammon-deevil
      On minds o' men,

But at the same time write fu' ceevil,
　　　　An' no offen'.

*　　*　　*　　*　　*　　*

My een look laigh!—here comes some earl,
To his ee-glaiss gien mony a twirl,
Fegs but he is a bumptious carle
　　　　Like noble ranks,
'Mang upper ten he is a pearl
　　　　For blackguard pranks.

His only boast's his pedigree,
Or some ancestor's chivalry,
(The polished name for robbery
　　　　O' sons o' toil,)
Whase richt was micht; sae they made free
　　　　Wi' English soil.

Wi' robbin' Will they first cam owre,
An' greatest thief had greatest power,
To rob, an' steal, an' a' devour
　　　　Wi' Norman fashion;

Titles an' dignities, a shower,
   Was Will's kent passion.

Frae sire to son, the lan' descends,
But this youth comes with vicious ends,
Whas name ower country sune extends,
   For warst o' ways,
Nae lordly virtue in him blends
   Wi' lordly praise.

He leads a life o' lazy leisure
Wi' bad examples withoot measure,
To satiate his lustful pleasure
   Begins gey sune ;
The squand'ring o' come-lichtly treasure—
   It's easy dune.

Weel marked wi' pleasure's vilest rounds,
Horse-racin', gamblin', queans, an' hounds,
Or shootin' birds on Heelan' grounds,
   A sad recital ;

Nought to his credit e'er redounds
　　　To gild his title.

Rank blin's the cost o' sic like fun,
Fast life hae noo inroads begun,
An' coffered wealth is seen to run
　　　In steady flow,
He, like puir fouk, whan it is dune,
　　　To pawn maun go.

Sic lives aft shine in country's page,
As them wha keep,—aff some mortgage—
Servan's an' coortly equipage
　　　Wi' ball and routs,
Wi' social nymphs fu' aft engage
　　　In drucken bouts.

A coronet can aften hide
Some deeds that couldna daylicht bide,
Nae streaks o' gude,—to see wi' pride,
　　　Their names adorn.

Gin men in them their hopes confide
> They're clipt or shorn.

Ay! thus puir farmers' rents are raised,
To keep up titled scamps—half crazed,
'Tis thus nobility 's debased,
> In nation's een,
An' revolutions aft hae blazed
> Against them keen.

Hoo lang will vile distinctions last?
Will gulf o' rank by nane be passed?
Will wealth, that's frae the soil amassed,
> Be spent by few,
In carryin' on the war o' caste?
> They yet may rue.

Nae nation yet has shown records
To prove the use o' men ca'd Lords,
Or titled chiels, whas life accords
> Wi' blackest evil;

Wealth in their han's fu sair affords
       Food for the deevil.

Would they but wield the power they hae,
Wi' doin' gude, in puir fouk's way,
How mony hearts wad for them pray,
       Their kindness touchin';
The puir man's prayer—wi' brichtest ray,—
       Adorns escutcheon.

Aye! gin their cash was in trade's use,
Free frae the " mak a' " vile abuse,
The ootlay profits wad produce,—
       Pairt they could spen'
In keepin' toilers frae Puir Hoose,
       Their days to en'.

The blessin's o' the gratefu' poor
Wad licht the path o' deein' oor,
The gude they've dune be entered sure
       In heavenly pages;

Their titled name wad lang endure,
      Thro' mony ages.

     *   *   *   *   *   *

Wha's this that crawls ower Fancy's stage,
Glib i' th' gab, wi' look fu sage?
He's ane that's ca'd, "Man o' th' Age,"
      An' is weel kent;
He lang-syne manhood did mortgage,
      To cent per cent.

Ten years hae no gane ower his heid
Sin' he was toilin' for his breid;
But noo, thro' cunnin' tricks indeed,
      He cocks his nose.
His ledger honest hearts wad bleed,
      An' tell o' woes.

Here stan's a man, wi' upcast heid,
Wha likes weel to be flunkeyèd,

He'd look gey laigh, gin debts war paid
        An' boo for ever ;
He kens the law hoo to evade,
        Being unco clever.

His gift o' gab, aft faith instils,
Saft men wi' prospects aft he fills,
O' roarin' business, free frae ills,
        Or dubious tricks.
Suspectin' not, they tak his bills,
        An' then get—nix.

He eats the corn, an' gie's the chaff,
To creditors, wi' chucklin' laugh ;
For them he maks this epitaph,
        Wi' mental glee ;
" Here lies the feckless human calf,
        Wha trustit me."

Mark weel, this wrecker o' his race,
Wha's impudence is a' his grace,

He shows his still unblushin' face,
        In public street ;
'Mang heartless men he is the ace,
        O' base deceit.

But aff my stage he noo is sneakin',
The Deil awaits, he is him seekin',
An' quick to place that aye is reekin',
        He'll sune be ta'en ;
Rascals like thae will get a smeekin',
        Intae the bane.

    *   *   *   *   *   *

Across the stage ane comes in view,
O' men like him I've seen a few,
Wha's cash wis sma' when first he knew,
        Employin' ways ;
By puir-paid men, his stock sune grew,
        Sae did his praise.

As cash increased, sae did his greed,
He aye preached poor, was aye in need,
His wife an' weans he'd barely feed,
      Or claes be givin',
Gold ! Gold ! his thocht, 'twas a' his creed,
      The siller savin'.

Were men designed to toil an' slave,
An' end their days in pauper's grave,
For men like him, wha live to save,
      Their ill-paid sweat,
Wha's only motto is " To Have,"
      Wi' siller great.

Awa ! wi' men whas weel-filled purse,
Sits sair upon them as a curse,
The way it's got is aftimes worse,
      Than doonricht stealin',
Their gowd, the god, to bring remorse,
      Whan life is failin'.

Hoo mony men wi' senseless clamor,

Wha hardly ken the English grammar,

Hae been shoved up wi' fortune's rammer,

      Frae midden heid ;

'Mang gentle fouk their gowden glamour,

      Stan's them in need.

Sometimes they will a trifle gie,

To ony wark o' charitie,

But mark ! when gi'en ! the're sure to flee,

      Wi' fussy capers,

An' spread their generositie

      In daily papers.

Aft to the Kirk, ye'll see them gau'in,

Wi' solemn face, an' meekly crawin',

As if the Lord's wark they war dae'in',

      Wi' liberal han's ;

Sabbath's the day whan they are sawin'

      Their grabbin' plans.

Oh why should life to men be given,
Wha's only god is siller savin',
Wha never think o' puir fouk's cravin',
      Or hunger's tale,
Whaur Winter wants rise aft to Heaven
        Wi' mournfu' wail ?

Will days e'er come whan wealth will be,
Content to live on humbler fee,
Whan man to man in harmony,
      Shall happy dwell,
Not rich—but proof to poverty,
        Hooe'er sae snell.

A man that's aftimes maist deservin',
Is some rich maister's ill-paid servan',
His hard-wrocht wage, lives ill preservin',
      In meat an' claes ;
Hoo can he sma' fund be reservin',
        For life's dark days ?

December win's may blaw gey cauld,
Doon reekless lums sough unco bauld,
The crouchin' puir a tale unfauld,
   Few, few will care ;
Rags, hunger, dirt, an' scenes untauld,
   Will at ye stare.

Will siller men e'er be content,
Or thankfu' live on less per cent,
Will Christian hearts ne'er gie consent,
   To help anither ;
Go ! see the puir, an' ye'll relent,
   An' save a brither.

The Auld, Auld Book, will aye record,
Wealth wasna made for men to hoard,
But guid to a' it should afford,
   An' life impart,
Close-handit wealth is aft a sword,
   To pierce the heart.

Come, men o' means, here is your field,
Whaur sma' ootlay is sure to yield
Per-centage great, aboon be sealed,
      Wi' joyful debit ;
A sma' accoont in Heaven revealed,
      Is to your credit.

Fortune may on ye favours shower,
But mark ! there comes a dreaded hour,
Ye may "will" then, but ah ! the power
      Is frae ye ta'en ;
Your gold-bound heart Death will devour,
      Wi' scathin' pain.

Whan he staps in wi' gentle nips,
Afore he gi'es ye harder grips,
As owre your een comes dark eclipse,
      Hear your last word,
My Gold ! my Gold ! the shiverin' lips
      Faintly record,

Ye glide awa, sae does your fame,
An' lea' behind a hollow name,
That sune dies oot, nane can proclaim,
      Ae lovin' spot ;
A life wi' selfish, sordid aim,
      Is sune forgot.

    *   *   *   *   *   *

Anither comes, wi' gentle mien,
Wha's noble deeds to a' are seen,
In helpin' puir fouk, wha hae been
      O' life's strings scant;
An' sendin' comfort to a wheen,
      In realms of want.

Come forrit man ! an' jink na there,
The likes o' ye is unco rare,
Ye ken fu weel your wealth to share,
      In guid resolves ;
Time's rust will nae thy name ootwear,
      While warl' revolves.

K

The sodger loon, or titled cad,

Has been by nations' homage clad,

An' toons hae aft gane statue mad

    Him to adore ;

For him wha mak's hearts unco glad,

    Sic 's nae in store.

The fechtin' chiel, for fechtin' laws,

Wha wi' a gleed ba' may be fa's,

For what is ca'd " oor country's cause,"

    Is made a hero !

His name will ring lang thro' Time's ha's,

    Anither Nero.

Are deeds o' bluid the noble base

O' fame, or boastfu' nations' praise ?

Are men wha kill the maist o' faes

    To be ca'd gude ?

The monuments to them ye raise,

    Are streaked wi' bluid.

Does fame consist in God's lives takin'?
Or will it shine wi' puir hearts breakin',
Or yet in thoosan's left forsaken,
        To starve or **die?**
This gildit fame, o' sic a makin',
        Wha would **envie?**

There is a fame that few discern,
An' maist **to a' gi'**es sma' concern;
Few, few its alphabet will learn,
        Or ways hae socht;
That Highest Fame, few barely **earn,**
        Or gi'et a thocht.

All **preachin'** eloquence is vain,
To whitewash England's foulest stain;
The Court o' Poverty will reign,
        Owre a' **supreme;**
A Christian's prestige few **maintain,**
        Fouk to redeem.

Here stan's a man, wi' business thrivin',
In him ye see nae conscience rivin',
Or framin' plans wi' sly contrivin',
　　　　To tak in a' ;
An' mak his " mair " wi' fouk deprivin',
　　　　O' hainin's sma'.

Ay ! tho' his gowd increased wi' years,
He thocht o' cauld want's bitter tears,
Wi' Christian heart he them uprears,
　　　　Wi' charity ;
Frae mony he's chased hunger's fears ;
　　　　Great rarity.

An' mony a widowed heart has prayed,
That blessin's wad rest on his head,
For a' the licht that he has made
　　　　In humble hames ;
His generous soul has lang displayed,
　　　　Love's kindest flames,

Thou benefactor o' your race,
Withoot aristocratic grace,
Nae bronze or stane may show your face
      In city's square ;
Ne'er mind—there is for ye a place
      In mansions fair.

His life o' gude, whan nigh ootspent,
Bears heaven's impress, whaur a' is kent,
He lea's the warl' wi' calm content,
      An' look divine ;
His gude name stan's his monument,
      That lang will shine.

Death comes to him wi' kindly smile,
Divested o' a' terrors vile,
An' lifts him gently owre the stile,
      To ither lan's ;
Attendants blest the change beguile
      Wi' heavenly sangs.

Waes me! my Muse (a cutty deft),
Wi' queenly curtsey, has me left;
O' her fond smile, I'm aince mair reft,
      An' numbers terse;
A web o' words withoot her weft
      Maks unco verse.

To rhymin' thochts I hae gi'en vent,
The M S to ye I hae sent,
An' Beattie, gin its shawn in print,
      'T may kittle creeds
O' hearts that harder are than flint;
      Fu o' misdeeds.

Wha reads an' thinks I've dune amiss,
In notin' fouk wi' siller bliss,
Contrite I'll stan' an' bauldly kiss,
      My ain richt han':
Wha angry gets at screed like this,
      Is no a man.

WEE WAVES

Aft, aft in the evenin' by Roker I've strayed,
An' watched the wee waves a' wi' moonbeams arrayed,
Come joyously curlin', an' singin' wad fa',
Fedeckin' the sands wi' an edgin' like snaw.
Ye wee waves aye dancin' what is it ye sing?
Do ye to the lane heart some glad tidings bring?
It canna be sorrow that maks ye sae free,
Wi' love ye are loupin' wee waves o' the sea.

>   Wee waves curlin', singin',
>   Loupin' blithe an' free,
>   Tales o' them aye bringin'
>   Sailin' on the sea.

Say, say hae ye always as onward ye flow,
Smiled sweetly on ane, that sae fondly I loe,
Or hae ye in anger whan wild tempests flee,

Disturbed his sweet slumbers whan dreamin' o' me?
I fain wad believe that some tale ye could tell,
Or whisper some message o' love to mysel'.
That hame he is comin' frae far owre the sea,
I think the wee waves are aye tellin' to me.

      Wee waves curlin', singin', &c.

Why should I be weary whan ye bring me joy?
Why should anxious fearin' life's pleasure destroy?
I'll nae mair be dowie, or feckless an' wae,
But blithely will lilt noo thy heart-cheerin' lay.
Roll, roll in thy beauty, an' lave ye the shore,
May nae rufflin' breezes the bosom blaw o'er;
Sing ye aye as cheery, an' comfort aye gie,
To lane hearts a-watchin' wee waves o' the sea.

      Wee waves curlin', singin', &c.

SCOTTISH HILLS

Hie thee away to our Scottish hills,
   And mist-capped mountains blue,
Her lakes and dells, with glowing fells,
   And vales of every hue.
The mountain steep, or black gorge deep,
   With awe our bosom fills ;
The heathery air, we breathe so rare,
   Belongs to Scottish hills.
      Hie thee away, &c.

The heather-bell we all love so well,
   So tender 'tis in form,
With bosom bare to the mountain air,
   It welcomes mountain storm.

And on rocky crags the antler'd stags,
   Free as the dancing rills,
Look with delight to the heath-clad height,
   And claim the Scottish hills.
     Hie thee away, &c.

THE WAUKRIFE WEAN

OH ! sleep my bairnie, hush my lammie,
　　This greetin' maks me wae,
Nae sleep comes to your ain mammy,
　　Your fidgin' tires me sae.
Ye canna tell me what is wrang,
　　Or gin ye are in pain,
I maunna think that ye are thrawn,
　　Altho' a waukrife wean.
　　　　Hushie ba, lammie loo,
　　　　Gang to sleep, my bonnie doo.

I carena tho' nae sleep I hae,
　　But for my guidman's sake,
Wha toils sae hard for us a' day,
　　I'd no like him to wake.

An' tho' I've hardly closed an ee',
  For twa-three nichts noo gane,
Wi' a' your care ye'r dear to me,
  Altho' a waukrife wean,
    Hushie ba, lammie loo,
    Gang to sleep, my bonnie doo.

Sae gang to sleep noo, hushie ba',
  An' cuddle in my bosie,
I'll saftly dicht the tears awa,
  An' mak ye snug an' cosie.
My joy an' pride's my bairnie braw—
  A toom hoose whar there's nane,
I'd want my sleep, aye meat an' a',
  For my ain waukrife wean.
    Hushie ba, lammie loo,
    Gang to sleep, my bonnie doo.

Tho' cares an' troubles bairnies gie,
  An' hearts wi' sorrows fill,

A mither's love wi' joy can see
    A pleasure in her toil.
An' when auld age comes owre us baith,
    At oor fireside alane,
The son may cheer life's doonward path,
    Tho' noo a waukrife wean.
        Hushie ba, lammie loo,
        He's gane to sleep, my bonnie doo.

FISHERMAN'S SONG

Away! away! from out of the bay,
   We leave the land behind;
Our craft sails well, with its brown lugsail,
   Filled with a flowing wind.
To the fishing ground o'er the sea we bound,
   And quickly cleave our way.
'Tis our delight, by the clear moonlight,
   To sing our cares away.
      With the rest of the fleet we gaily meet,
         Hoist sail and scud away,
      O'er the heaving sea we all bound so free,
         And welcome coming day.

Our brown nets long we shoot with a song,
   While shining herrings sweep,

O'er the sea we mark their lighted track,
    As flying through the deep.
We strongly pull, when each mesh is full,
    Of silvery creatures bright,
And fill our boat, as we gaily float,
    While stars bid us good night.
        With the rest of the fleet, &c.

Our sail we trim by the daylight dim,
    The craft seems to be free,
As her bosom laves the morning waves,
    For home we cleave the sea.
And we near the land where loved ones stand
    To welcome us again ;
With our heavy take, all for their sake,
    Hard won upon the main.
        With the rest of the fleet, &c.

AVAUNT! DESPONDENCY

Avaunt! Despondency, thy clouds shall not preside
  O'er my fair firmament of constant day;
No! Gloomy Witherer! my heart, whate'er betide,
  Shall be invulnerable to thy sway,
And smiling, fling thy brooding wiles away.
  Unworthy, pitiable man, whom all deride
Is he, who kindling in himself thy fire
  Binds his poor being in its funeral pyre;
Shrunk is his worth, outraged is manhood's pride,
  As blindly, moodily, he courts his mind's decay.

Free be my soul to soar; let it love's impress bear;
  No vile antithesis of life be mine.
Happy in childhood I; why should the manhood wear
  A leaden, mummied look, or moping whine
At Fortune's changes? No, never shall I repine!

But dauntless will confront calamities and tear
From out the ruin broader Happiness,
   Then calmly think on earthly nothingness,
Thanking my God that I have reaped a share
   Of stern enlightenment; and deem th' exchange
     divine.

So I will cherish life e'en in adversity,
   Hope at the prow, Contentment at the wheel.
He makes his little less who feeds despondency,
   And self-deprived is void of earthly weal;
Unmanned, his disappointments keener feel.
   Behold I stand beneath the heaven's canopy,
Vast as itself, love ruleth over all;
   But man, poor man, whom trifles will enthrall
Excludes love's light, whene'er despondent he,
   And warps Hope's attributes that aye would joys
     reveal.

NEVER FASH YOUR THOOM

Fouk shouldna be doonhearted wi'
    The ups an' doons o' life ;
Fouk shouldna ither fouk envie
    In claes, in cash, or wife.
Whate'er oor lot in life may be,
    Gin spade, or plough, or loom,
Be aye content wi' what you hae,
    An' never fash yer thoom.
        Sae sail thro' life wi' happy mind,
           Tho' ithers faster soom ;
        Ye needna fret tho' left behind—
           Hoots, never fash yer thoom !

The win's, ye ken, blaw here an' there,
    Sometimes wi' calm or gale ;

An' sun, tho' hid, some rays can spare,
    But whaur we canna' tell.
Whan Nature's sides are dark an' bricht,
    It shaws fouk's heads are toom,
Gin they e'er think a' should be licht,
    An' needless fash their thoom.
        Sae sail tho' life wi' happy mind, &c.

Fouk aye should live heart-thankfu' wi'
    Their lot hooe'er its placed,
The mind o' man shines bonnilie,
    Whan wi' content 'tis graced.
We a' upon ae level lie
    Whan in the wee cauld room,
The en' will come—the best will die—
    Sae never fash yer thoom.
        But sail thro' life wi' happy mind, &c.

'TIS THE CAULD WINTRY WIN'S
A-SIGHIN'

'Tis a twelve-months to-day since your faither gaed
  Oot wi' his ship to the wild, wild sea ;
'Tis a twelve-months to-day since my heart was dead,
  Yet I live, my dear laddie, for thee.
I see his braw face in your ain an' rejoice,
  But the look aye sets me a-cryin',
An' I aften think that I hear his loved voice
  In the cauld wintry win's a-sighin'.
     'Tis the cauld wintry win's a-sighin',
      'Tis the moan o' the sea I hear,
     An' I think on your faither dyin',
      Far at sea, an' no loved anes near.

I ne'er can forget his last smile or guidbye,
  His pairtin' aye brings me mair grief ;

Cauld noo is oor cot, I may greet, I may sigh,
   There is naething to gie me relief ;
I gaze on the sea, an' I watch the waves rise,
   Aboon me the snaw-clouds are flyin',
An' I feel that my heart is like cheerless skies,
   Wi' the cauld wintry win's a-sighin'.
       'Tis the cauld wintry win's a-sighin, &c.

My faitherless laddie, your mither is wae,
   The sun o' her life is fast settin',
I pray aft to Him wha's the puir widow's stay,
   That my bairn he'll no be forgettin'.
The wintry win's whisper, as wildly they rave,
   That soon 'neath the sod I'll be lyin',
Awa frae a' cares I'll find rest in the grave,
   An' owre me the win's will be sighin'.
       'Tis the cauld wintry win's a-sighin', &c.

TEENIE'S GRAVE

THERE is a lone sequestered spot,
  Where memory fondly clings ;
The longing heart forgets it not,
  And silent weeping brings.
The ruined church still stands, I see,
  And willows still do wave ;
The Tay still sings its lullaby,
  Beside my Teenie's grave.

    Flow on, sing on, thou lovely Tay,
      Ye night winds sigh and rave ;
    Ye willows nod to ruins grey,
      Around my Teenie's grave !

My buried love, I tread once more
  The quiet ground and holy,
I see the grass deep growing o'er

Thy hallowed mound so lowly.
Cease beating heart! flow burning tears!
  Life's sole relief I crave;
Grant me, dread Power, who prayer hears,
  My wish on Teenie's grave.
    Flow on, sing on, thou lovely Tay, &c.

I live, but am with life decoyed;
  To Teenie I would fly.
My life is blank —all, all is void—
  I would that I could die;   .
For freedom to my grief-bound life,
  Death's darkest hour I'd brave.
Oh! happy moment if this strife,
  Would end o'er Teenie's grave.
    Flow on, sing on, thou lovely Tay, &c.

I see her form, I hear her voice.
  She beckons me away
To lands of bliss where all rejoice,

And loving souls have sway.
Fain would I come : oh, hear my prayer,
   Ye who life to me gave ;
Take back the life—I would be there
   With Teenie in the grave.
     Flow on, sing on, thou lovely Tay,
       Ye night winds sigh and rave ;
     Ye willows nod to ruins grey,
       O'er mine and Teenie's grave !

MY MUSE

WHAN toilin' hard the lee-lang day,
  My Muse will whisp'rin' tell,
"Just sing your sangs, whate'er fouk say,
  They're pleasin' to mysel'."

I CANNA sing in polished lays,
  Some lang Arthurian theme ;
My Muse, bein' clad in hamespun claes,
  Will unco bare aye seem.
I canna boast o' college powers,
  Nor words o' livin' fire :
I've dwelt in nae braw rose-clad bowers,
  An' tuned nae gowden lyre.

I hae nae name for deeds o' worth,
  Nor hae I riches' wiles ;

I ken nae great fouk o' the earth,
 To flourish 'neath their smiles.
I ne'er hae trod in noble ha',
 Whar gowd is a' the licht,
Nor picked the crumbs that sometimes fa'
 To flatt'ry-pourin' wicht.

I ne'er hae daubed wi' sleekit rhyme
 Prood fouk o' high degree,
Nor hae I sang in measured chime
 O' worth, whar nane should be.
I ne'er wad beck an' boo to chiels
 Wha's gowd was a' their base.
The mind fu' sma' aft gen'rous feels
 Whan paintin' it wi' praise.

Tho' fouth o' gowd aft mak's fouk great,
 It ne'er can gie them sense;
Lea' sic the sovereigns wi' their state,
 Lea' me my sangs wi' pence.

Sae I will sing o' humbler hames,
   An' tread in worthier fields :
Unslavish Fancy's nobler aims
   Me purest pleasure yields.

In cosie hoose, graced wi' my Jean,
   An' share o' warl's gear,
I coort my Muse an' ne'er compleen—
   Cauld poortith I ne'er fear.
I'll thrum no mercenary lay
   Where sordid passions dwell,
But sing my sangs, whate'er fouk say,
   They're pleasin' to mysel'.

DINNA TAK THE RUE

Jamie.   OH ! lea' me not, I've loo'ed ye lang,

           My heart ye ken is thine,

      The dearest thou, them a' amang,

           Say noo that ye'll be mine.

      I canna thole the thocht ava,

           Hoots Mary, 'tis na true,

      To lichtly throw my love awa,

           An' say ye've ta'en the rue.

                Fu weel ye ken you're a' to me,

                    An' weel ye ken I loe,

                Nane but yersel' 'mang a' I see,

                    Sae " Dinna tak the rue."

Mary.   Man Jamie, dinna deave me sae,

           Ye dinna ken my heart ;

      To hae to say't, in troth I'm wae,

           But you an' I maun part.

To days gane by, to a' ye've said,
　To love's hame fancy drew,
I bid fareweel, an' winna wed,
　For I hae ta'en the rue.
　　Some keekin' glints o' happiness,
　　May yet my life imbue,
　　An' Jamie, ye may thankfu' bless
　　The day I took the rue.

Jamie.　Had I e'er dreamt this day wad come,
　Whan first I saw your ee,
Langsyne I'd gane ahint the drum,
　An' left my a' an' thee.
I've little bindin' noo to hame,
　O' freens I hae but few ;
Life noo has tint its only aim,
　Sin' ye hae ta'en the rue.

But Mary whan I'm far frae thee,
　Nae sorrow may ye feel ;
My foremost feelin' aye shall be

To wish the lassie weel.
A bitter draught is severed love,
　My heart wi' pain is **fou'** ;
Mair bitter still my life shall prove,
　Sin' ye hae ta'en the rue.

　　Ae pairtin' kiss, an' then fareweel,
　　Nae mair I'll seek to woo,
　　Unkent to a', I'll fain conceal,
　　That some ane took the rue.

Mary.　**Ha! ha**! my Jamie, dinna grieve,
　　Hope lifts again **its broo,**
　You're a' to me,—an' wi' your leave,
　I'll tak anither rue.

Jamie.　By Powers aboon! by a' that's dear,
　　Come to my **arms my** doo ;
　Then a' is fause, love's bands appear,
　Far stronger o' a rue.

SCRAP CASTINGS

NOT UNCOMMON.

SOME folk are Fortune's favored **weans,**
And in her lap are holden ;
Their copper penny **flung in air**
Falls down to them **a gold un.**

A WISH.

Fain wad I see a' mankind proof
To whisky's wiles, an' in their loof,
The fell desire wi' **grip o' steel**
Maist crush't afore its powers **they feel.**

AN EPITAPH.

Here lies a man o' business fame,
Whase life had only **ae** intent,
To lea' behind the hollow name
O' havin' made his cent per cent.

AN OPINION.

In takin' or in gi'en a dram,
I see nae harm, but whan fouk cram
Glass efter glass adown their throat
They spoil the gude ane micht hae wrought.

THE LAST NAME.

Time's catafalque was borne along,
By angels from some heavenly City,
From withered world, moved in the throng
One mourner lone, his name was B——e.

TO J——N R——E.

R——e, remember, fractions make up wholes,
Consume all ashes, save the owner's coals,
A master good has he, who hath a D——n,
A servant true have they, who have you R——e.

A SNOB.

Cease ! flatterer, cease thy hollow ways,
Think not to please me with thy praise ;

Send your congratulations oft,
But prithee! do not write so soft;
What tho' you act the serpent part,
No skin can hide your craven heart;
Mean, soulless acts, you ne'er will lack,
You'd stab your friend behind his back.
Your highest feelings, as ink spot,
Evaporate, and leave their blot;
Wise ignorance, on you bestowed,
Supremely hath its blest abode;
On manly years, child-wisdom dawning,
Developed is, when thou art fawning.

A VERSE.

Bright hopes that fill the heart,
Oft on life's rocks are dashed;
And rays that joy impart,
Are oft too bright to last.

There's some fouk leevin' in this warl,
Wha butter baith sides o' a farl.

M

Gold may be good, but Honesty is best,
Stick to the latter, Time will do the rest.

Wealth without servants' toil and trouble
Would soon leave masters in a hobble.

Fouk wha buy things they dinna need
Aft pay dear for their buyin' greed.

Wha spends a shillin' to hain a penny
Aft thinks he's savin' wi' his money.

Gold in the pouch, nocht in the heid,
Maks a queer gentleman indeed.

Sons aft are spoiled wi' fortins ready made,
Take aff their cash—they starve withoot a trade

Some fouk wi' fortins an' attendant trains
Hae made the same wi' stealin' ithers' brains.

The best o' men you e'er can find
Are those wha to your fauts are blind.

Life, is the engine all mankind doth wear,
Death is its stoppage for Divine repair.

The baby, youth, and man,—a grand machine,—
Could chance such perfect movements e'er convene?
Out fool! no more thy vagaries uprear,
Or doubt the being of the Engineer.

DEBT.

Chief o' a' ills that stain the human race,
Wrecker o' happiness, wi' keenest barb,
Ye bring het blushes to the conscious face,
An' shroud the mind in hideous gloomy garb.
Great Tempter! into thy embrace fouk fa',
Kennin' fu' weel ye are a bitter curse;
Rather than live content on unco sma',

Oh happy thief, who e'er can pilfer,
That rural gem called M——y T——r,
Convicted he—but, know ye this,
His punishment is earthly bliss.

A STRANGE SUICIDE

WHILE sauntering one eve upon the pier,
Whose stony bulwarks bind the sportive Wear ;
Before me shone the great mysterious sea,
Humming its low but peaceful melody ;
Behind, the land, faced with its scenic coast,
A land to love, meet for its dwellers' boast.
I met a man whose melancholy mien,
Accorded ill with such a glorious scene,
      Unequalled scene.
From deepening blue above, presaging night,
Some wak'ning stars gave forth their feeble light,
And dancing waves below with ceaseless song
Their revels held the silv'ry beach along.
Seaward afar, I marked the glancing swell,
Of ocean's bosom, as it rose and fell ;
With fav'ring breeze on to some distant shore,

Full hundred sail-clad ships it proudly bore,

Whose whitened sails shone in departing day,

And clung with fondness to each narrowing ray.

Anon, amid this throng so haughtily,

The collier-steamers passed them swiftly by,

Heedless of tides or quarter of the wind,

Leaving their spiral smoky trails behind.

Nearer in shore their tiny boats to fill,

The patient fishers plied their lines with skill,

And luckless few whose efforts vain appear

Hoist their brown sail and shoreward songless steer.

E'en Roker cliffs loomed loftier in their shade,

And echoed sounds in wave-washed caverns made.

Westward, loved Tunstall reared its head so brown,

The silent warder o'er the busy town,

Whose muffled hum, on night-breeze borne along,

Proclaimed life's echoes from its teeming throng ;

Upwards o'er all, the lamps' reflections tinge

The night-clouds grey, that hang with crimson fringe.

Southward, but dim, the pinnacles of trade

And twinkling lights of Seaham port displayed ;

Seaham! thou guerdon of Vane's wealth and will,

Beholding thee, my wish is Onward still ;

Scenes such as these my pen can ill convey

In words their beauty, or their worth pourtray.

I met a man :

Accosting him, " Good e'en," I said,

He blankly looked, and shook his head—

Methought him dull—louder, " Good e'en,

Hast ever viewed so fair a scene ? "

His eyes beamed with peculiar glow,

And gazed above, around, below,

Then stared me full with brows close-knit,

While o'er his face there seemed to flit

Despairing smile, so sickly, wan.

Ere it had passed, he thus began.—

" Fair, fair without, dark, dark within,

So like those that this world dwell in."

One hand across his brow he drew,

The vacant gaze more vacant grew ;

The other, clenched, hung by his side,

He suddenly outstretchèd wide,
And mine he seized with clammy clasp :
Oh horror, 'twas a maniac's grasp !
He whispered low " Art thou my friend ?
If so, fear not, come to the end ;
We'll mark the loving, laughing waves
Their peans sing on drowned men's graves."
'Mid doubtings, yet to please his whim,
Prepared for aught, I walked with him
O'er slimy stones, clad with seaweed,
Requiring caution's utmost need ;
Till passing wave my feet had wet.
" Come now, no farther can we get,"
I said, then stood, e'en so did he.
Then gazing downwards wistfully,
" Not yet ! he said, Ha ! Ha ! No ! No !
You, friend, must first my life-tale know."
Thro' twilight dim, I marked the glare
Of eyes that seemed to say beware !
I marked the twitchings of his lip,
And nerved myself for treach'rous grip ;

He waved his hand, I watched each move,

Anticipating sudden shove.

" Ah me ! Ah me ! " he slowly said,

Then mournful calmness all o'erspread

His count'nance, which but beamed to show

One fleeting ray of Reason's glow ;

Then pointing up, " Will I dwell there,

Where all is love, where all is fair ?

Fair is this world, strange is its strife ;

Doth zone of love adorn thy life ?

Not so with me, Ha ! Ha ! a blank

My life hath been, bound with the rank

Weeds of self-love, that wither life

Of man who hath a loveless wife.

A loveless wife, ingrate and base,

Will rust the throne where Reason sways ;

'Tis true, tho' strange, her love is hate ;

Blame not the man when such his fate.

Love only from the mouth is such

That ne'er the heart of man can touch.

No comfort there, home cannot bring

Attachments fond, that else would cling.
I've toiled and spent her whims to please;
Her mighty self nought could appease.
Fox-like her cunning, soulless, venal;
My life hath been far worse than penal,
Yet, to her faults I have been blind,
Tried oft to woo her, being kind.
Ha! Ha! 'twas vain, it was no use,
No worthy point could I produce;
For years I've borne such life as this,
Oft sought in other scenes the bliss
That should have graced my own fireside,
But which to me has been denied."
He paused, and clearly I could trace
Hate's workings varnishing his face.
I pitied him, e'en met his views,
And tried by persuasive ruse
To draw him from the tempting sea.
'Twas vain, he only laughed at me.
He gazed above and then below,
And whisp'ring said, " Not yet, No! No!

The impulse ere long will prevail,
But friend, come listen to this tale.

" A woman, o'er whose brainless head
Some five and twenty summers fled ;
Despair was feeling, yet would oft,
As one at sixteen, blush as soft,
Had tried to woo each youth who passed,
Without her toils, till one at last
Was caught within, then fierce her grip,
And such a hold she ne'er let slip ;
This plastic youth, to her appears
In sense a boy, and boy in years.
He knew not women's ways or arts,
Or how they play deceitful parts,
Or how they duplex natures keep,
Till fixed the unsuspecting sheep ;
She, on her single prospect thought,
Still thinking on, this feeling wrought ;
—Die she would not a vile old maid,
Come young or old she would be wed.

Ay, he was soft, scarce seventeen turned ;
So like himself—the calf-love burned
Within his breast, when she began
With cunning wiles, him to trepan.
As dying miser grasps his gold,
As wolf prowls round the midnight fold,
As spider lurks some fly to catch,
She watched, and planned this youth to snatch.
One thought alone her vile mind carried,
By hook or crook she would be married ;
She would not wait his worth to scan,
'Twas all the same, if just a man.
Ha ! Ha ! she thought on better halves,
And well she knew young ones are calves "
He paused, and gave a hollow laugh,
" Ho ! Ho ! Ha ! Ha ! Oh what a calf !
Beginnings small oft make great ends,
Success oft sin and wrong attends.
'Twas so with her, sin-luck, forsooth,
Shone with her o'er this senseless youth ;
To gain her aim, with sly caprice,

She stooped to ev'ry false device ;
—" Her age was equal to his own,
Her love for him, how stronger grown ! "—
Oft would she take the golden course,
And hint she owned a well-filled purse,
Blinding the youth, who cleanly caged,
Thought her perfection, and engaged
To marry her, who so adored him,
Whose constant love would aye reward him.
Parental warnings got no heed,
As Jezebel could now him lead ;
Experienced hints he boldly parried,—
'Twas done ! She won ! The pair were married.
Some lives oft have an untold sequel,
When married they, with age unequal.
Marriage is blest when love abides,
And unity of hearts presides,
When consummated ne'er to find
Life's hidden shoals erst undefined,
That wreck the hopes, that peace destroy
Drive from the home the promised joy,

Their presence serving but to whet,
The dismal future of regret.
Yea, these she showed, ere yet a week,
This loving one who was so meek.
Ha! Ha! she proved her hollow heart;
And how she played the serpent's part!
Too late, he found when ope'd his eyes,
Protested love aye shallow lies.
Cared she for none, did as she chose,
Laid on the youth his life of woes;
As Time rolled on she aye got worse,
And proved indeed a bitter curse,
Her little sense got wondrous less;
Her only love—was love of dress.
Aimless, loveless, never thinking,
Often tippling,—(never drinking.)
Ho! Ho!
Home a mockery and vision,
Nought for him but her derision,
Self, concentrated in her pride,
Her ev'ry movement ne'er could hide.

Decade of life, in misery led,
He meekly bore till hope had fled ;
Within, around, now all is dead,
And ruined Reason reigns instead.
Ho ! Ho !" He fiercely clutched my hands,
" Before you now this wife-dupe stands !
All would be fair, did love abound,
Then happier world could ne'er be found ;
Weary it is, worthless, and void,
If peace of mind is so destroyed.
Long I for that rest-giving shore,—
The hope of life, when life is o'er :
Thus will I now its secrets reap."
We tugged,—I fell,—he took the leap ;
Loud rang his cry on ev'ning air
" All, all is fair ! All, all is fair."

With dreamy gaze I saw the waters close
O'er him, and watched, but ne'er again he rose.
Confusedly I marked the bubbling swirl
Of life's last efforts spread in circling whirl.

Knew not the gentle waves an unloosed soul
Had thro' their crests ascended to its goal ;
Still rolled they on, and in their ceaseless flow,
A death-chant sang o'er him who lay below.
The gathering night its curtain slowly spread,
As moodily I homeward journeyéd,
Communing with myself, my reasoning gave
Impressed belief, that happier in the grave
Is he, who battling outward, worldly strife,
Receives at home the gyves of loveless wife.
O bliss the best, chief of all wishes human,
The purest thou diffused by loving Woman.

LET A' BE BEHADDEN TO NANE

Hech! I am behadden to nane,
  I live, an' I've nae cause to fret ;
My pouch an' its a' is my ain,
  A fig for the chield that's in debt.
Wi' little I'm aye weel content,
  The mair, gin unpaid, I'll no hae ;
My honour is guid, I can want,
  An' soundly I sleep whan I'm sae.
      For I'll be behadden to nane,
        But aye independent will be ;
        A naething is he wha is fain, .
        On ither fouk's wings aye to flee.

I'll ne'er be wi' creditors teased,
  Na, Honesty lives wi' me yet ;
My wifie an' sel' are weel pleased,

Oor Freedom is cursed na wi' debt;
I carry my head unco high,
   Wi' conscience as bricht as the morn;
I nod to kent fouk passin' by,
   They canna look on me wi' scorn.
        For I am behadden to nane, &c.

I wadna live lives that I ken
   Tho' riches afore me you'd set,
I ne'er wad be marked among men
   As "Gentleman droonin' in debt."
Na, na, I've a pride in my heart,
   That rins wi' delicht in ilk vein;
'Tis—act aye the honest man's pairt
   An' want what ye canna maintain.
        Sae I'll be behadden to nane, &c.

Gleg spiders o' mankind appear,
   An' gild wi' deception their net,
They fatten on ither fouk's gear,
   An' live oot a life aye in debt.

Withoot,—a' is gabbin' an' show,
    Within,—they are strangers to worth,
Nae virtues wi' them e'er ootflow,
    For Honesty died at their birth.
            Sae I'll be behadden to nane, &c.

Awa wi' the men wha are blin',
    To Honesty's angel-made smile ;
Unhonoured 'mang kith an' 'mang kin,
    They flourish,—but 'tis to defile,—
Aff the earth they are snoovled awa,
    Nane speak a kind word o' regret ;
Ae freen is awaitin' their ca',
    Wha gars them pay weel for their debt.
        Lat a' be behadden to nane,
          An' shun wi' contempt fause support,
        Hae naething but what is your ain,
          Lat,—pay-as-ye-gang,—be your forte.

WAUKEN, MY SANG!

CAN Scotland nae mair roose the spirit that slumbers,
  Amang her blue mountains, her lakes, an' her dells?
An' is she bereft o' her ain lassie's numbers,
  Whase shadow aroun' her still ling'ringly dwells?
Na, na, there are whisp'rin's aince mair she is wakin'
  Frae oot o' the bower whaur lang she has lain;
An' aff frae her mantle the dew-draps are shakin',
  An' deckin' her broo wi' the heather again.

    Wauken, my sang! let me walcome her risin';
    Roose ye my soul, be wi' ecstasy fired;
    Shout, ye blue mountains, a heartfelt orison,
    An' be ye, auld Scotland, wi' homage inspired.
        Gie her walcome then,
        Lowlan' an' Hielan' men,
    Auld Scotland's sang-lassie has waukened
    agen.

She treads owre the fields whaur o' yore she did
　　wander,

　An' views wi' delicht her abodes o' lang syne ;

Owre hills an' thro' glens whaur loved rivers meander

　Is heard her joy-cry; "Bonnie Scotland, still mine."

She sees her braw dochters aye bonnie an' bloomin',

　An' marks her famed sons aye uphaudin' her worth;

She greets whan she sees them their virtues entombin',

　But smiles in her pride whan they lichten the earth.

　　Wauken, my sang! lat me welcome her risin', &c.

In ha' an' in cottage her sangs they are singin',

　An' aft at the winnocks she listens wi' glee ;

They gladden her heart, tho' aft to her upbringin',

　Sair thochts o' the past, an' the tear to her e'e.

She thinks on the sons wha but lived to adore her,

　An' laid at her feet gems o' untarnished sheen ;

An' sadly conjurin' their loved forms afore her,

　Wi' joys sees the laurels she gae them still green.

　　Wauken, my sang! lat me welcome her risin', &c.

She sighs as she sings in the sweetest o' voices,
  An' saft mountain breezes the echoes prolong;
"Ken ye, my lov'd Scotland, my spirit rejoices,
  Whan sons o' my choosin' extol thee in song?
Aft vot'ries fu' mony I've decked wi' the laurel,
  But thine in my heart warmest feelin's instil;
Nae nobler or brichter shines in my famed warl,—
  Their sangs Bonnie Scotland, gie me pleasure still."
    Wauken, my sang! lat me welcome her risin';
    Roose ye my soul, be wi' ecstasy fired;
    Shout, ye blue mountains, a heartfelt orison,
    An' be ye auld Scotland wi' homage inspired.
      Treat her gey kindly then,
      Lowlan' an' Hielan' men,
Auld Scotland's sang-lassie has waukened agen.

THE ROBIN

My bonnie wee robin, whaur hae ye come frae ?
    Whaur gat ye that sang ye are singin' the day ?
Wha tell't ye to come, ere the Autumn's awa,
    An' lilt oot your warnin' o' Nature's doonfa' ?
        Sing, my bonnie robin, sing,
           Walcome aye to me ;
        Winter's tidin's tho' ye bring,
           Pleasure great ye gie.

The broon leaves the trees an' the brackens adorn,
    An' sangs o' the reapers are heard in the morn ;
The swallows are dartin' on licht joyous wing,
    An' hummin' bees still to wee flowrets do cling.
        Sing, my bonnie robin, sing, &c.

Cam ye my wee birdie a walcome to gie
    To sunset o' summer still lingrin' a-wee ?

Your notes a' unechoed wi' sorrow are boun',
　Say, will ye lilt blither whan nicht fa's aroun'?
　　Sing, my bonnie robin, sing, &c.

Are ye my wee robin, foretellin o' wae?
　Is beauty to wither at thy saddenin' lay?
Are scenes noo enchantin' to lose a' their worth?
　Is nocht to be left but a snaw-mantled earth?
　　Sing, my bonnie robin, sing, &c.

Hoo strongly, lane warbler, ye tell to my heart,
　That beauties o' Nature an' life maun depart;
Ye walcome earth's winter wi' notes ringin' free,—
　I wish I could walcome my winter like thee.
　　Sing, my bonnie robin, sing, &c.

THE AULD SWORD

THE bugle sounds fu' lood an' clear,
　On Hielan' hearts its notes are fallin' ;
They heedna hame, nor a' that's dear,
　While Scotland is upon them callin'.
Gae bring the auld sword aince mair doon,
　There's valour in the grip o't dwellin' ;
My gran'sires focht for hame an' croon,
　Fu' doughtily the fae repellin'.
　　　Noo I maun gang awa frae thee,
　　　　See ! see ! braw lads the ca' obeyin',
　　　The banner's waving bonnilie,
　　　　A coward he, wha langer stayin'.

Bind ! bind the sword ! oh cease thy tears,
　I hear ! I hear the distant firin' ;
Thy love maun yield ; the fae appears,

Oor country's danger sons requirin',
The bugle sounds its last refrain,
    Fareweel! may angels be thee guardin'.
I'll bring the auld sword hame again,
    A mother's pride, my heart rewardin'.
        I'll ne'er disgrace the auld, auld blade,
        It's worth an' deeds my courage nervin';
        "Go! go my son, be round thy head,
        A mother's prayers danger swervin'."

THE HARVEST MOON

A SKETCH AT WHITESIDE FARM, TORTHORWALD

THE harvest full moon in the east gently keekit,
   Surroondit wi' dark cloods o' roseate hue,
Awa owre her path fleecy cloodlets war streakit,
   Like pearlins a' set in enamel o' blue.
Higher, up higher, the nicht queen was glidin',
   Till oot frae the cloods she in majesty shone ;
Sae modest an' tender, sae calmly presidin',
   She reigned in her beauty supremely alone.

The cloodlets abashed far afore her a' leapit,
   Till a' as an ocean o' love lay serene ;
Companions in beauty, the waukened stars peepit,
   Bestuddin' her throne wi' their joy twinklin' sheen.
Tho' a' gae her homage o' heavenly fulness,
   It seemed that her rays tinged wi' sorrow were shed,

Nae welcomes or joy-shouts disturbed the death still-
    ness,
  An' maist a' unheedit she smiled on owrehead.

The fields, in their twilight, aroon' me war glowin',
  Tho' auld age was bendin' their yellow taps doon ;
The wimplin' bit burnie fu' near me was flowin',
  A siller thread twined in a garment o' broon.
Afar in the valley the sweet Nith was glancin',
  Like sunbeams at play with the gems on a croon,
A' burnished wi' beauty, ilk wavelet was dancin',
  An' singin' its praise to the full Harvest Moon.

Enrapt was my soul wi' the prospect afore me,
  Tho' close to my heart clung a feelin' o' wae,
An' prayin', I wished, whan life's sunset cam o'er me,
  That calm as this nicht be the end o' that day.
Surely the hame o' thae rays saftly beamin',
  Is free frae the sorrows that bind this life roon,
That aye in thae regions pure love ever teemin',
  Is whisp'ringly told by the full Harvest Moon.

IMPRÒMPTUS

WRITTEN ON VISITING THE GLOBE TAVERN, DUMFRIES.

TIS said !

" Earth's gorgeous palaces and cloud capped towers,

With mighty monuments upreared among,

Yea ; the Great Globe, by superhuman powers

Shall be dissolved, and all to chaos flung."

But yet :—

In this dire wreck one atom on thro' space

Shall fly intact, and ever on career,

Bearing this Globe, whose walls beheld the face

Of Burns, whose genius oft did revel here.

Till then :—

Tho' sleeping well in yonder worthy tomb,

Bared be thy head ! on entering this room.

And Scots !—

So long as Criffel rears its reverend head,

So long as Nith flows winding o'er its bed,

So long as Scotsmen ne'er to wrongs succumb,

Be here our Mecca for all time to come !

LINES

Written on observing Burns' Chair so sacrilegiously cut and hacked
by Visitor's Initials, &c.

Stranger ! be this your heartfelt prayer,

As in this hallowed nook ye sit,

That ——— be he who cuts the chair

Whereon oft flashed our Burns' wit ;

Unmanly he, and diabolical,

Meet native of a place symbolical.

OOR AIN SEL'S TWA

FEGS ! we've been married fifty years, an' mony fechts
   we've seen,
  In trachlin' thro' the muir o' life, wi' a' its blasts
   atween ;
Oor bairnies hae a' grown to men, an' frae us gaen awa,
  An' nane sit by the fireside noo, but oor ain sel's twa.

An' Meggie, as I think upon the years we've spent
   thegither,
  A feelin' grows that closer draws oor hearts to ane
   anither,
There's something waukens in my breist I canna tell
   ava,
  I'm owrecome wi' the memories o' oor ain sel's twa.

I aft reca' the days o' youth, sae fu' o' love an' bliss,

An' whan I paintit fancied years, o' earthly happiness
I had nae thocht but hoo to serve, the brawest o' the
     braw,
     Love whispered that the warl had nane, but oor ain
          sel's twa.

We've felt fu' aft, mid flicht o' Time, adversities a
     wheen,
     In dark or licht, my Meggie lass, ye proved my
          truest freen ;
Tho' fause anes aye wad turn their backs whan we
     were at the wa',
     There was some sterlin' comfort clung, to oor ain
          sel's twa.

We baith hae toiled frae morn to nicht, to keep the
     hoosie haill,
     An' bring oor bairnies up that they the richt should
          never fail ;
An' mony a tear ye shed for them whan siller was but
     sma',

But somehoo blessins restit aye on oor ain sel's twa.

Sae years hae flown, the hoose is toom, whaur a' aince
    rang wi' mirth,
  Some fecht the warl for themsel, an' some are in the
    earth !
Aft owre my cheek the tears o' thocht will silently
    doon fa',
  For them whas lauch aince cheered the hearts o'
    oor ain sel's twa.

We've borne o' sorrows great oor share, tho' sometimes
    tastin' joys,
  The warst has been for gude to us, in makin' baith
    mair wise ;
Oor journey's en' is no far aff, sae in oor Faither's ha'
  Unpairtit there, I hope will be aye oor ain sel's twa.

Oor fifty years o' weddit life, hae knit love's bands sae
    close,

To sever them—hoo sair the thocht!—ane couldna
    bear the loss,
A longin' wish I hae that aye, my heart encircles a',
    That on ae grave the grass may wave owre oor ain
        sel's twa.

WEE WATTIE

YOU'RE no a mither's bairn,
You're sic a steerie callan,
O' wark she gets her sairin',
You're just anither Allan ;
I fear we've ca'd ye wrang,
If sae it is a pity,
You're gettin' unco thrawn
An' no like Wattie Beattie.
> Oh the bairn, the tiresome laddie,
> Sae gang awa to your ain daddie.

Ye pu' the dishes o'er,
An' like to see them brakin',
Whan they fa' on the floor,
You're sure to fa' a-lauchin' ;
Ye'll no sleep soun' at nicht,

But will hae oot your feetie,
Ye craw whan it is licht
An' no like Wattie Beattie.
    Oh my bairn, my steerin' laddie,
      Tak him a wee, I'm tired noo, daddie.

Ye rumlie, prattlin' chiel,
Ye've in ye the richt mettle,
To fecht the warl weel,
An' Fortune sure to fettle.
Nae sma' man ye may be,
Tho' noo ye pu' the teetie,
Wha lives will in ye see
A worthy Wattie Beattie.
    Loup my bairn, my roarin' callan,
      You're like your faither, Willie Allan.

www.ingramcontent.com/pod-product-compliance
Lightning Source LLC
Chambersburg PA
CBHW021703210326
41599CB00013B/1498

* 9 7 8 3 7 4 4 7 3 0 6 9 3 *